3D Printing and Sustainable Product Development

The text focuses on the role and the importance of 3D printing in new product development processes. It covers various aspects such as the 3D printing revolution and Industry 4.0, sustainability and 3D printing, and economics of 3D printing. It discusses important concepts, including 3D printing, rapid prototyping, mechanical and physical properties of 3D printed parts, nanomaterials, and material aspects of 3D printing.

Features

1. Presents recent advances such as Industry 4.0, 4D printing, 3D material mechanical characterization, and printing of advanced materials.
2. Highlights the interdisciplinary aspects of 3D printing, particularly in biomedical, and aerospace engineering.
3. Discusses mechanical and physical properties of 3D printed parts, material aspects, and process parameters.
4. Showcases topics such as rapid prototyping, medical equipment design, and biomimetics related to the role of 3D printing in new product development.
5. Covers applications of 3D printing in diverse areas, including automotive, aerospace engineering, medical, and marine industry.

It will serve as an ideal reference text for senior undergraduate, graduate students, and researchers in diverse engineering domains, including manufacturing, mechanical, aerospace, automotive, and industrial.

3D Printing and Sustainable Product Development

Edited by
Mir Irfan Ul Haq
Ankush Raina
Nida Naveed

CRC Press
Taylor & Francis Group
Boca Raton London New York

CRC Press is an imprint of the
Taylor & Francis Group, an **informa** business

Designed cover image: © shutterstock

First edition published 2024
by CRC Press
6000 Broken Sound Parkway NW, Suite 300, Boca Raton, FL 33487-2742

and by CRC Press
4 Park Square, Milton Park, Abingdon, Oxon, OX14 4RN

CRC Press is an imprint of Taylor & Francis Group, LLC

© 2024 selection and editorial matter, Mir Irfan Ul Haq, Ankush Raina and Nida Naveed individual chapters, the contributors

Library of Congress Cataloguing-in-Publication Data
Names: Ul Haq, Mir Irfan, editor. | Raina, Ankush, editor. | Naveed, Nida, editor.
Title: 3D printing and sustainable product development / edited by Mir Irfan Ul Haq, Ankush Raina, Nida Naveed.
Other titles: Three D printing and sustainable product development Description: First edition. | Boca Raton, FL : CRC Press, 2023. | Includes bibliographical references and index. |
Summary: "The text focuses on the role and the importance of 3D printing in new product development processes. It covers various aspects such as the 3D printing revolution and Industry 4.0, sustainability and 3D printing, and economics of 3D printing. It discusses important concepts including 3D printing, rapid prototyping, mechanical and physical properties of 3D printed parts, nanomaterials, and material aspects of 3D printing. The book presents recent advances such as industry 4.0, 4D printing, 3D material mechanical characterization, and printing of advanced materials and highlights the interdisciplinary aspects of 3D printing particularly in biomedical, and aerospace engineering. It discusses mechanical and physical properties of 3D printed parts, material aspects, and process parameters while showcasing topics such as rapid prototyping, medical equipment design, and biomimetics related to the role of 3D printing in new product development. It also covers applications of 3D printing in diverse areas including automotive, aerospace engineering, medical, and marine industry. It will serve as an ideal reference text for senior undergraduate, graduate students, and researchers in diverse engineering domains including manufacturing, mechanical, aerospace, automotive, and industrial"-- Provided by publisher.
Identifiers: LCCN 2023011471 (print) | LCCN 2023011472 (ebook) | ISBN 9781032306803 (hbk) | ISBN 9781032306827 (pbk) | ISBN 9781003306238 (ebk)
Subjects: LCSH: Additive manufacturing. | Three-dimensional printing. | Green products--Design and construction. | Rapid prototyping.
Classification: LCC TS183.25 .A15 2023 (print) | LCC TS183.25 (ebook) | DDC 621.9/880286--dc23/eng/20230506
LC record available at https://lccn.loc.gov/2023011471
LC ebook record available at https://lccn.loc.gov/2023011472

ISBN: 978-1-032-30680-3 (hbk)
ISBN: 978-1-032-30682-7 (pbk)
ISBN: 978-1-003-30623-8 (ebk)

DOI: 10.1201/9781003306238

Typeset in Sabon
by MPS Limited, Dehradun

Contents

Preface

With increasing competition in the global market and the increased consciousness amongst the end users about environment, industries throughout the globe are competing to deliver new products with improved efficiency and performance. Further, there is a growing demand for shortening the product development time and delivering low-cost and light-weight products having improved performance, and at the same time, the products need to be designed in a way that they can be recycled and lead to zero waste to minimize their environmental impact and truly assist the sustainable cradle-to-cradle approach. With the advancements in the field of additive manufacturing, 3D printing has evolved as an alternative to subtractive manufacturing and offers benefits such as low material wastage, ability to handle part complexity, less human intervention and low cost. Therefore, 3D printing can play a vital role in new product development by delivering products at faster speed and reducing energy consumption, thereby leading to significant cost savings. Moreover, this technology has lot of potential to emerge as a sustainable manufacturing technology and can be useful to reduce production of scrap waste by reusing and recycling 3D printed parts.

The book is focused to highlight the role 3D printing can play in new product development. Different aspects of 3D printing such as sustainability, safety and recycling have been touched in various chapters of this book. Special focus has been laid to present different application areas of 3D printing such as medical and nanotechnology. A special chapter has been focused on providing an introduction to 4D printing. In addition to the written matter, the book contains supporting schematic diagrams, data tables and illustrations for easy understanding of the readers. Apart from providing state of art details on various material aspects of 3D printing, the book also highlights the various opportunities and challenges associated with the technology.

The features of the book appeal to both students and to researchers finding newer and better solutions to a lot of problems faced during new product development. The editors feel that the book shall serve the purpose of creating awareness among all stakeholders related to additive manufacturing industry so as to develop products which are cost effective and greener in short time.

About the Editors

Mir Irfan Ul Haq is currently working as assistant professor at the School of Mechanical Engineering, Shri Mata Vaishno Devi University (SMDV). He has previously worked with the R&D wing of Mahindra and Mahindra. He has obtained bachelor of technology degree in mechanical engineering from SMVD University and master of engineering in mechanical system design (Gold Medalist) from National Institute of Technology Srinagar. He has carried his Ph.D. in the area of tribology of lightweight materials. He is actively involved in teaching and research for the past ten years in the field of materials, tribology and 3D printing. His research interests include lightweight materials, new product development and additive manufacturing, development of green lubricants, self-lubricating materials and cutting fluids, wear of materials, and surface engineering. He has published around 50 research papers in SCI and Scopus indexed journals. He has edited three books and has served as guest editor in reputed journals. Moreover, he has been awarded various research grants for various projects from various agencies like DST and NPIU AICTE. He has attended numerous conferences and workshops both in India and abroad like the USA and Singapore. He is a member of reviewer boards of various international journals apart from chairing technical sessions in various international and national conferences. Dr. Haq is actively involved in organizing various conferences and workshops. He has also coordinated various student events such as SAEBAJA, ECOKART and TEDX. Moreover, he has supervised around 20 projects on master's and bachelor's level. Dr. Haq has served as member of various committees at national and state levels.

Ankush Raina is an assistant professor in the School of Mechanical Engineering, Shri Mata Vaishno Devi University, Jammu and Kashmir, India. His area of interest includes wear and lubrication, additive manufacturing, and rheological properties of lubricating oils. He has been awarded Ph.D. in the area of industrial lubrication. He completed his M. Tech in Mechanical System Design from NIT Srinagar, Jammu and Kashmir. He was awarded with Gold Medal for securing first position in M. Tech. He has extensive research experience in the field of nano lubrication and has carried out various studies using nano-additives in different lubricating oils and composite materials. Further, he has also explored the tribological characteristics of different types of 3D printed polymeric materials. He has published more than 50 articles in the SCI/SCIE/Scopus indexed journals with more than 1,200 citations and h-index of 19 and i-10 index of 32. He has also presented several papers in national and international conferences in India and abroad. He is the reviewer of several reputed SCI/SCIE indexed journals and has also coordinated several events at university level.

Nida Naveed is currently a senior lecturer and programme leader at University of Sunderland, UK, and has been working in academia and industry for more than fifteen years. Before joining the University of Sunderland, she has worked with NCG Newcastle College Group, UK, as a programme leader in engineering and at The Open University (OU), UK, as an honorary associate (researcher) at the Department of Engineering and Innovation. Previously, she was also a Research Candidate at The Open University for MPhil leading to PhD on a fully funded studentship by Rolls Royce Limited, East Midlands Development Agency (EMDA) and OU. She has obtained her PhD in materials engineering from the Department of Engineering and Innovation at The Open University, UK. She received her master's (MEng) in structural engineering and bachelor's (BEng) in civil engineering from NED Engineering University. Dr. Nida has a Postgraduate Certificate of Education (PGCE) in mathematics and numeracy specialists from Teesside University, UK. She is a chartered engineer and a member of the Institution of Engineering and Technology (MIET), UK. She is also a fellow of the Higher Education Academy (FHEA) UK. Dr. Nida has published various articles in reputed journals and is currently guiding various master's and PhD students. She is a member of Athena Swan Team for the Faculty of Technology, University of Sunderland, UK. She is a mom, and enjoys cycling and swimming.

Contributors

Mark Armstrong
University of Sunderland
United Kingdom

Naveen Kumar Bankapalli
Indian Institute of Technology
    Mandi
India

Mahdi Bodaghi
Nottingham Trent University
Nottingham, United Kingdom

L. Bazli
Iran University of Science and
    Technology
Iran

W.Y. Chak
Hong Kong Metropolitan
    University
Hong Kong

Deepak Chhabra
M.D.U., Rohtak
India

Eddy Chan
Hong Kong Metropolitan
    University
Hong Kong

A. Esmaeilkhanian
Amirkabir University of
    Technology
Iran

Ramesh Kumar Garg
D.C.R.U.S.T.
Murthal, India

Sumit Gahletia
D.C.R.U.S.T.
Murthal, India

Vishal Gupta
Indian Institute of Technology
    Mandi
Himachal Pradesh, India

Mir Irfan Ul Haq
Shri Mata Vaishno Devi University
Katra, Jammu and Kashmir, India

Ashish Kaushik
D.C.R.U.S.T.
Murthal, India

S. Khaksar
The University of Georgia
USA

W.K. Kwong
Hong Kong Metropolitan
    University
Hong Kong

C.C. Lee
Hong Kong Metropolitan
    University
Hong Kong

T.T. Lee
Hong Kong Metropolitan
   University
Hong Kong

C.H. Li
Hong Kong Metropolitan
   University
Hong Kong

S.L. Mak
Hong Kong Metropolitan
   University
Hong Kong

Hamid Mehrabi
University of Sunderland
United Kingdom

Marwan Nafea
University of Nottingham Malaysia
Malaysia

Nida Naveed
University of Sunderland
United Kingdom

Özgür Poyraz
Eskisehir Technical University
Turkiye

Upender Punia
D.C.R.U.S.T.
Murthal, India

Ankush Raina
Shri Mata Vaishno Devi University
Katra, Jammu and Kashmir, India

Asrar Rafiq Bhat
Indian Institute of Technology
   Mandi
Himachal Pradesh, India

M. Reisi Nafchi
Islamic Azad University Najafabad
   Branch
Iran

F. Sadeghi
Islamic Azad University of Yazd
Iran

Binnur Sagbas
Yildiz Technical University
Turkiye

Anmol Sharma
Guru Gobind Singh Indraprastha
   University
India

F. Sharifianjazi
The University of Georgia
USA

Prateek Saxena
Indian Institute of Technology
   Mandi
Himachal Pradesh, India

W.F. Tang
Hong Kong Metropolitan
   University
Hong Kong

Haishang Wu
University of Sunderland
United Kingdom

M.Y. Wu
Hong Kong Metropolitan
   University, Hong Kong

Ali Zolfagharian
Deakin University
Australia

Chapter 1

3D Printing and New Product Development

Opportunities and Challenges

Asrar Rafiq Bhat, Vishal Gupta, Naveen Kumar Bankapalli, Prateek Saxena, Ankush Raina, and Mir Irfan Ul Haq

CONTENTS

1.1 INTRODUCTION TO ADDITIVE MANUFACTURING

Additive manufacturing, commonly known as three-dimensional (3D) printing is a process, where objects or components of desired 3D CAD models are built by depositing materials layer by layer (illustrated in Figure 1.1). In contrary to conventional manufacturing processes where parts are made by removing unwanted material from a block of raw material, there is no need for cutting tools in additive manufacturing so, this also called as tool less method of manufacturing. Additive manufacturing provides a free hand for designing the complex structure of best performance by removing design limitations of subtractive manufacturing techniques (Bajpai et al., 2020; Saxena et al., 2021). A wide range of metals, food materials, biomaterials, composites, alloys, building materials, smart materials,

DOI: 10.1201/9781003306238-1

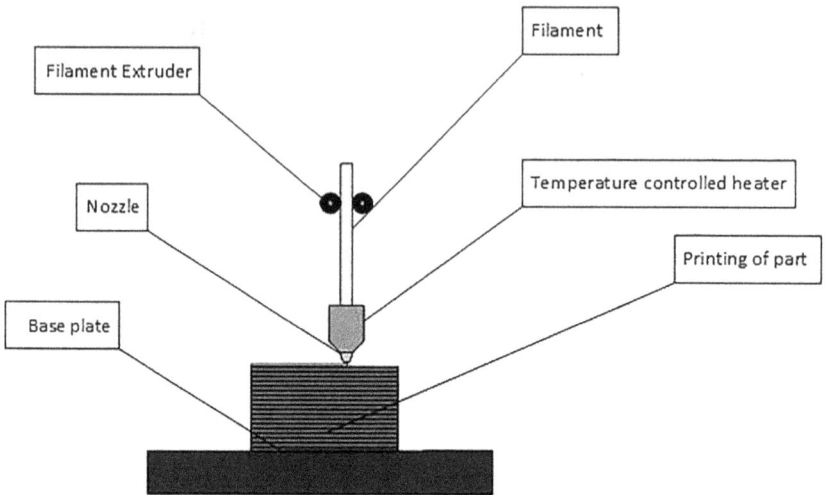

Figure 1.1 Schematic 3D printing using material extrusion process (FDM).

polymers, and ceramics can be 3D printed using different additive manufacturing techniques (Bhatia and Sehgal, 2021).

AM is used to make lightweight and cost-saving components for automotive and aviation applications and many more advanced applications, such as bio-printing of tissues, sensors, integration of electronics and construction industry (as represented in Figure 1.2 (Vaezi et al., 2013; Jandyal et al., 2022). Additive manufacturing is the best technology to fabricate components made from expensive materials like nickel and titanium because of less wastage of material in the layer upon layer addition process, and due to this, additive manufacturing caught the attention of academia and industry, mainly aerospace industry (Malik et al., 2022). The additive manufacturing process was previously used extensively for prototyping of new product development, but nowadays parts made by 3D printing or additive manufacturing are used directly in the final product and sometimes 3D printed parts are used as spare parts (Rouf et al., 2022a; Chaturvedi et al., 2022).

The most widely used 3D printing technologies currently available are powder bed fusion, wire arc additive manufacturing, direct energy deposition, and sheet lamination, which are primarily used for metals. For plastics, stereolithography apparatus (SLA), material extrusion (FDM), and material

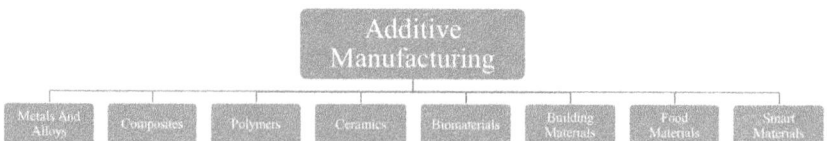

Figure 1.2 Materials used in 3D printing.

jetting processes are used. Direct ink writing technologies, powder bed fusion, and fused deposition modelling (FDM) procedures, as well as selective laser melting (SLM) and selective laser sintering (SLS), are employed for ceramics extrusion. Nowadays, a process combining extrusion-based 3D printing, followed by debinding and sintering, is used to manufacture metals and ceramics. In this process, filament containing polymer is reinforced with powder of the required metal or ceramic elements. The chapter's following parts include a brief discussion of a few of these processes.

Three-dimensional printing techniques are further classified into Micro and Macro fabrication. Different micro additive manufacturing processes are represented in the flow chart (Figure 1.3). When the size of fabricating object is less than 100 μm, that process is the microfabrication, and if the size is above 100 μm, that process is a macro fabrication. Miniaturization of products increases the demand for micro components in biotechnology, electronics, energy, medical, communications, automotive, and optics (Bajpai et al., 2020). But at the microscale, manufacturing complex structures using standard fabrication techniques is complicated than macroscale.

Additive manufacturing plays a vital role in removing design limitations to realize complex components with the best performance. Three-dimensional Micro-AM can be classified into three main groups, viz. 3D direct writing technologies, Scalable micro-AM systems, and Hybrid Processes. Microfabrication by AM can be realized for a wide range of materials like polymers, metals, ceramics, and composites (Bajpai et al., 2020). However, the majority of micro-AM techniques are still in the research stage. Micro stereolithography and electrochemical fabrication techniques among micro-AM systems showed more promising outcomes in 3D microfabrication when compared to other techniques (Vaezi et al., 2013).

In recent years, 3D printing technology has been evolving into four-dimensional (4D) printing by adding the fourth dimension, which is time (Figure 1.4). That is, in 4D printing, 3D printed objects change their shape over time or in response to an environmental stimulus (Hirt et al., 2017).

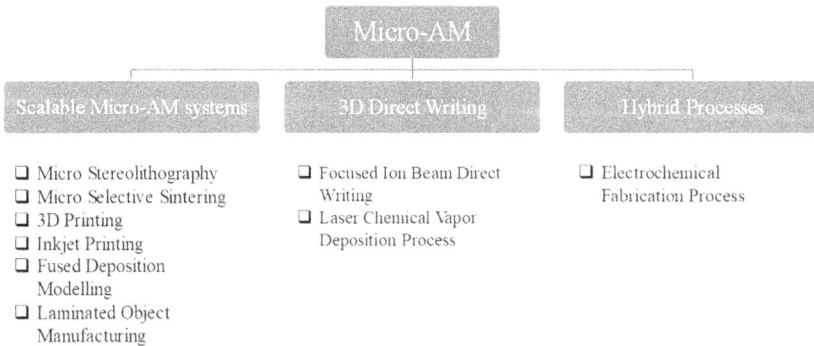

Figure 1.3 Micro additive manufacturing processes.

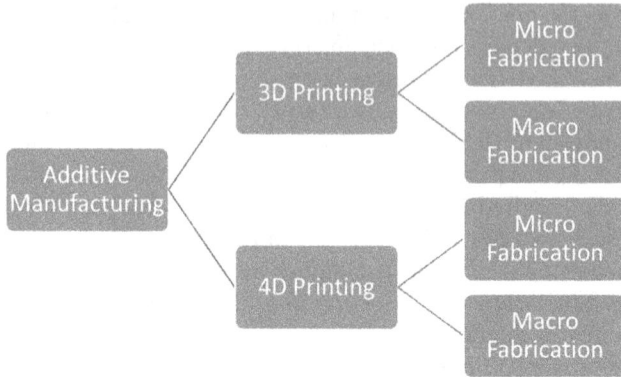

Figure 1.4 Classification of micro and macro in 3D and 4D printing.

In 4D printing, the objects are printed using 3D printed techniques, but the change in morphology is due to the material used in 3D printing. Smart materials like shape memory composites, shape memory polymers, and shape memory alloys are used to change the morphology of 3D objects. And shape change occurs when 3D objects are exposed to external stimuli like heat, light, ultrasound, pH, solvents, electricity, magnetism, etc. Potential applications of the 4D printing technique are in biomedical, like drug delivery, tissue engineering, implant organ, skin reconstruction, and bone reconstruction.

1.1.1 History of additive manufacturing

The first technique of additive manufacturing was used for fabricating models and prototypes in the late 1980s, but after 40 years of development, Nowadays, AM technology is among the most rapidly developing technologies in the world. During this development period, processing of wide range materials from the low melting point polymer materials to the high melting point ceramics and metals made feasible through the AM process. Initially used additive manufacturing processes like stereolithography apparatus (SLA), fused deposition modeling (FDM), selective laser sintering (SLS), and laminated object manufacturing (LOM) were typically used for fabricating of prototypes from polymers.

Apart from the processing of different materials, materials of various forms also made possible through AM techniques like materials in the form of liquid, powder, and wire. AM technology for fusible materials using a precursor of 3D laser cladding established in the 1971 patent of ciraud. In the patent of householder, the idea of making a 3D part using the molding process with the help of SLS systems is proposed. Due to high priced laser systems and unavailability of powerful computers at that time, these technologies were not commercialized. The first deterministic turning machine, which is also called DTM, was developed by the University of Texas and used this device as SLS, which entered the market as Sinterisation 2000/2500 machines along with powder materials like Rapid steel, DTM Rapid Tool.

In 1994, after acquiring patents from 3D Systems and the University of Texas, EOS GmbH Electro-optical systems introduced laser sintering machines to fabricated prototypes of plastics, which is called EOSINT P350. Later in 1995, the same company introduced a direct metal laser sintering machine (DMLS) to make plastic injection molding tools based on additive manufacturing process called EOSINT M250. This is the start of rapid tooling. In the same year, Fraunhofer Institute for Laser Technology (ILT) in Aachen, Germany, started research on the selective laser sintering process and achieved German patent in the pioneering phase. Some of the researchers and GmbH worked together on this technology. After all these improvements and inventions, accuracy and resolution have improved, and prices of the machine also decreased. This technology is playing a pivotal role in the fabrication of complex parts with better properties than before.

1.1.2 Working of an additive manufacturing process

In general, different steps involved in various 3D printing processes are discussed below:

1.1.2.1 Creation of the CAD model of the part to be printed

The part of being 3D printed created using a computer aided design software package. To get accurate results, it is better to use a solid modeling system rather than a wireframe model.

1.1.2.2 Generation of STL file

The input file for all additive manufacturing processes is STL (stereolithographic, it is the first AM technique). As different software packages use different algorithms for representing a CAD file. So, to establish consistency, STL format standardized in the 3D printing industries. In STL file, 3D model is described in triangles.

1.1.2.3 Slicing

Generated STL file is sliced into thin layers while slicing the location of the model to print, the orientation of model, and layer thickness; also, we adjust. According to size, it is essential to decide the location of printing, and orientation determines the time required to build. Also, many software packages allow the users to give supports needed for the object to be printed, which can be done before slicing.

1.1.2.4 Printing

Actual part building happens in this step using a fixed AM process. From polymer, metal alloy, the ceramic machine creates one layer at a time; the cross-section of each layer represents the sliced layer of the model. Depending on the type of materials, machine, the printing parameters change from machine to machine.

1.1.2.5 Post-processing

This step involves removing the part from build plate or machine and re-moving supports given. Some components may require cleaning-polishing, and apart from cleaning and polishing, some metallic parts require heat treatment.

This chapter covers numerous 3D printing technologies and different aspects of 3D printing. Material aspects of these 3D printing techniques have also been discussed. This chapter also covers the challenges associated with 3D printing in various applications for product realization, as well as the usage of 3D printing in new product development and other opportunities across engineering disciplines. Three-dimensional printing applications in the fields of aerospace, automotive, biomedicine, and electronics, as well as the individual parts utilized in each, has also been discussed at the end of this chapter.

1.2 VARIOUS 3D PRINTING TECHNOLOGIES

Three-dimensional printing is used as the synonyms for the all AM processes but in actuality all the methods vary depending upon the raw materials used, their processing such as deposition method and the energy source utilized. To standardize the processes, in year 2010, American Society for Testing and Materials (ASTM) F42 committee classified the AM processes in seven main categories. These categories are as shown in Figure 1.5.

For a type of material, it is possible to use different process as per the requirement. For example, printing of polymer component can be done with the help of the VAT photo-polymerization, material extrusion and powder bed fusion. Similarly, for metal 3D printing, powder bed fusion, material extrusion as well as direct energy deposition can be used. In the

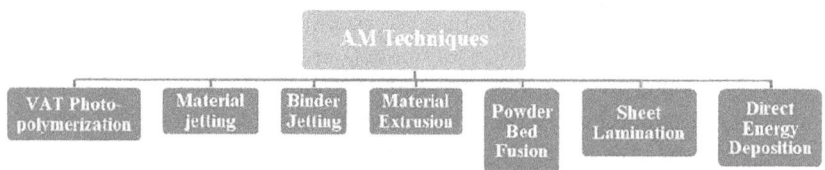

Figure 1.5 Classification of AM techniques according to ASTM F42 Committee.

subsequent sub headings, different type of AM techniques will be discussed in details.

1.2.1 **VAT photo-polymerization**

The stereolithography (SLA) method is another name for VAT photo-polymerization. It was developed in 1986 and is one of the early AM techniques (Melchels et al., 2010). The schematic illustration of SLA is shown in Figure 1.6(a). In this method, a layer of resin or monomer solution is applied with liquid thermoset polymer resins that can be cured with UV light (or an electron beam) to start a chain reaction (Saxena et al., 2020; Halloran, 2016). After being activated by UV light, the monomers quickly transform into polymer chains (radicalization). Following polymerization, a pattern within the resin layer solidifies to hold the following layers in place. After printing is finished, the unreacted resin is removed. The SLA has two different approaches: a top-down approach and a bottom-up approach. Some printed parts may undergo a post-processing procedure to get the desired mechanical performance, such as heating or photo-curing.

1.2.2 Powder bed fusion

Powder bed fusion based additive manufacturing is one of the important technology which utilizes high-power laser light as an energy source to build parts by selectively melting of metal/polymer powders in a layer by layer as given in the STL file {illustrated in Figure 1.6(b)}. To get parts without defects, some of the vital process parameters should be tuned carefully. The critical parameters which affect mechanical properties of fabricated parts are laser scan speed, hatch style, hatch overlaps, laser power, and hatch distance. The entire process should take place in an oxygen-free environment. This process was considered as versatile additive manufacturing process because of its ability to process a wide variety of materials, which include Fe-alloys, Cu- alloys, Al-alloys, Ni-alloys, Ti-alloys, Co-alloys, polymers, and their composites (Sutton et al., 2017; Gorji et al., 2020). In this process, we can get the required mechanical properties of materials by varying the process parameter.

1.2.3 Binder jetting

Binder jetting is one of the technique of AM that prints the object in layer by layer fashion by depositing the two type of material; first one is the build material and second one is the binder material. Most often the building material used to be in the powder form and the adhesive material is in the liquid phase (Gonzalez et al., 2016; Sivarupan et al., 2021). In this technique, binder is utilized as adhesive material to bond the layer of powder spread over the build platform or previously deposited layer as shown in

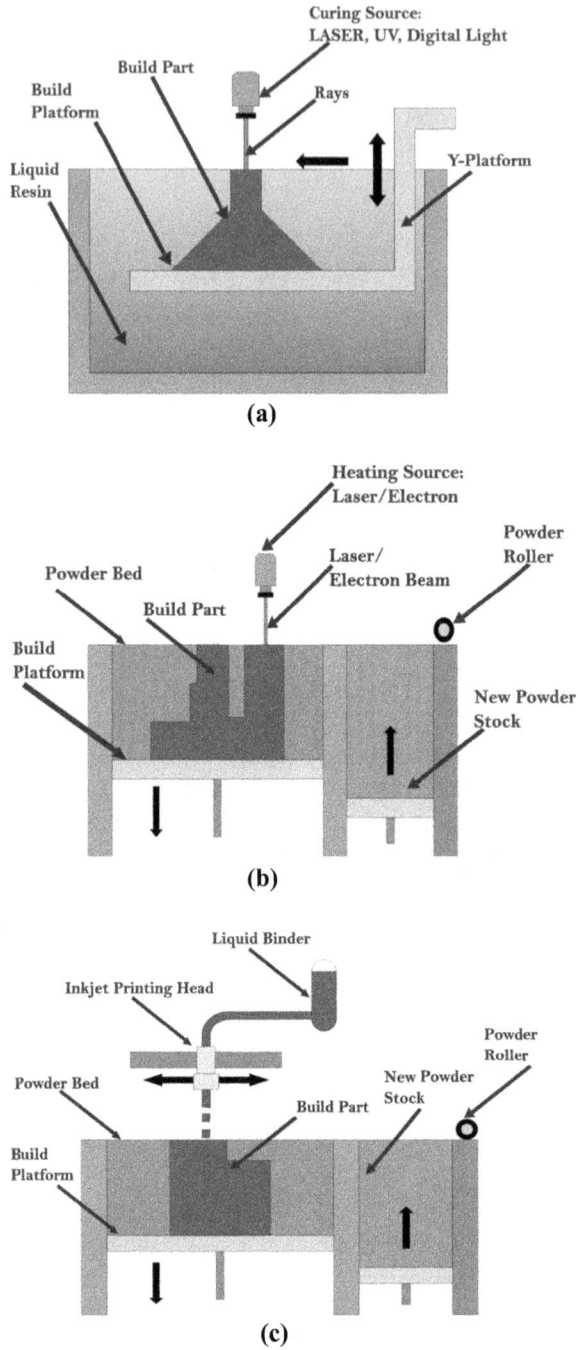

Figure 1.6 Schematics of (a) VAT photo-polymerization (b) Powder bed fusion (c) Binder jetting.

Source: Conceptualized from Bajpai et al. (2020).

Figure 1.6(c). A binder jetting setup have a print head that moves horizontally for the deposition of the binder liquid and a wiper mechanism is also present that spread the powder layer of predefined thickness over the build platform. After printing every layer, the build platform goes down in order of the layer thickness and process repeated. The part performance of the 3D printed components with binder jetting is not optimal to use it as the structural part due to its binding procedure which increases the processing time despite of being relatively fast during binder jetting. Unlike material extrusion, the support structure is not required in this technique.

1.2.4 Material extrusion

Material extrusion is one of the most widely used technique for the 3D printing of the end user components. Material extrusion is also known as the fused filament fabrication method. This technique was earlier used for the 3D printing of the polymeric components. Nowadays advancement is material extrusion process allows for the printing of metal and ceramic materials as well as the polymer matrix composite manufacturing. In this technique a continuous polymer filament is fed into the heated nozzle where the filament gets melted to semi solid state and extruded through the nozzle as shown in Figure 1.7(a) (Peng et al., 2018). This extruded material is deposited onto the previously printed layer. During printing, the layer that being extruded fuses into previously printed layer. The printing speed, layer thickness, layer width, infill pattern, and orientation of the layer affect the 3D printed part performance. The key advantages of FDM are its low price, high speed, and ease of use. The main drawbacks of FDM, on the other hand, are limited thermoplastic materials and poor surface quality.

1.2.5 Sheet lamination

Sheet lamination technique of additive manufacturing is more or less similar to traditional composite manufacturing. In this techniques, raw material to be used is in the form of the sheets that are bonded with each other by the means of the adhesive or with the help of welding {illustrated in Figure 1.7(b)}. Hence, sheet lamination is mainly divided into two parts; laminated object manufacturing (LOM) and ultrasonic additive manufacturing (UAM) (Derazkola et al., 2020). In UAM process the sheets or ribbons of the metal is joined with each other by the help of welding caused by the ultrasonic machines. Similar to traditional manufacturing, this process also needs the post processing steps like use of CNC machines to remove the unbounded metals. In LOM processing, the strips is of the paper material and glued them together to form the object. To make the post-build, simple cross hatching is used during printing process. Laminated objects are frequently utilized for aesthetic and visual models and are not appropriate for structural use. UAM uses metals like titanium, stainless steel, copper, and aluminum.

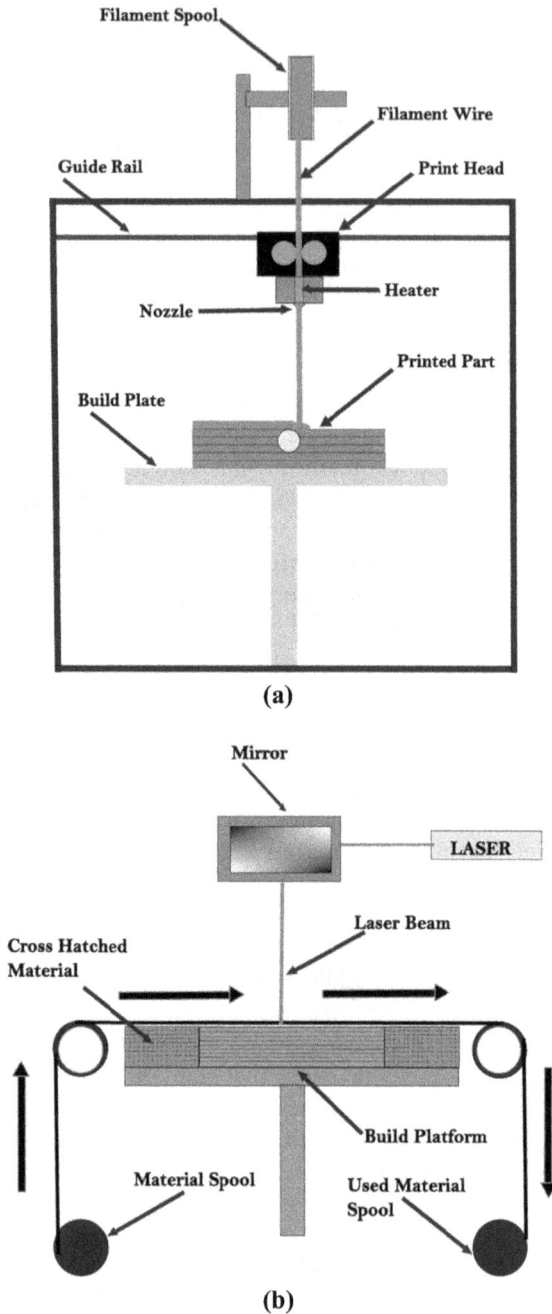

Figure 1.7 Schematics of (a) Material extrusion (FDM) (b) Sheet lamination.

Source: Conceptualized from Bajpai et al. (2020).

Internal geometries can be generated thanks to the low temperature of the procedure. The method connects a range of materials and consumes relatively little energy because the metal is not heated.

1.2.6 Direct energy deposition

Directed energy deposition (DED) includes different processes such as "laser designed net shaping," "directed light fabrication," "direct metal deposition," and "3D laser cladding." This is the most complex printing technology that is used to repair or increase the strength of the existing part (Shim et al., 2016). A schematic representation of DED is shown in Figure 1.8(a) and (b). A DED setup consists of the nozzle similar to material extrusion that is fixed on the robotic arm that can move in many directions. Metal used in this setup is in the form of the wire or powder (Gibson et al., 2021). Apart from the metals, polymer and ceramics can also be processed in this. The molten material is deposited on the specified surface where it gets hardened. The material can be placed from any angle and is melted upon deposition with a laser or electron beam owing to 4 and 5 axis CNC machines. Typical applications include fixing and maintaining structural parts.

1.2.7 Material jetting

Material jetting is an AM technique that works similar to two-dimensional (2D) inkjet printer. Material jetting is done in two ways in this technique. It may be either Drop on Demand (DoD) or jetting of material continuously on the build platform (Sing et al., 2020). Object in this method is built up layer by layer on the build platform or on the previously deposited layer by spraying and solidifying the raw material as illustrated in Figure 1.8(c).

In general, material is deposited by the help of the nozzle attached to a print head that moves across the build platform horizontally. Material is deposited by a nozzle that moves across the build platform horizontally. The deposited layer is hardened by curing with UV lights in case of the UV curable polymer resin. Material deposited on the part or build platform must be properly controlled as it will decide the thickness of the layer and hence performance and aesthetics of the parts produced by this technology. Polymers and waxes are useful and often used materials because of their viscous nature and ability to form droplets. Table 1.1 presents a comparative of various technologies.

1.3 MATERIAL ASPECTS OF 3D PRINTING

The next concern that comes to mind after going through many forms of 3D printing is what are the various materials that are utilized in these processes,

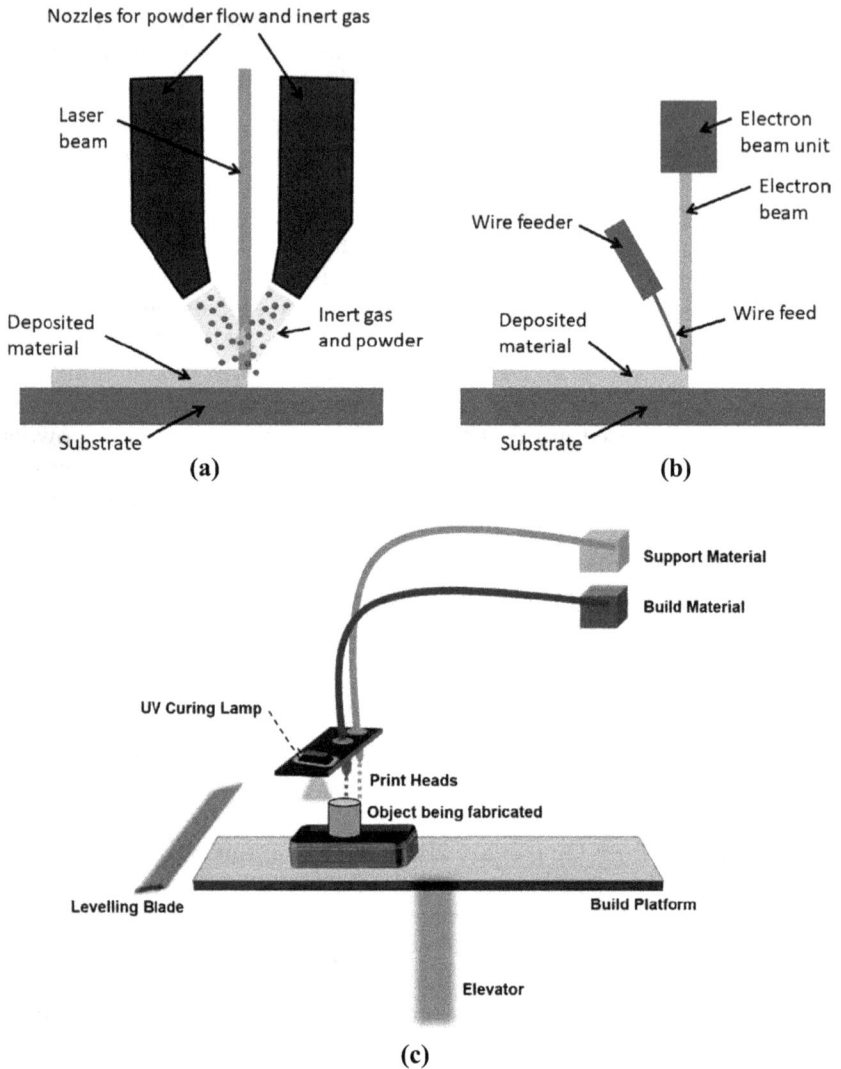

Figure 1.8 Schematics of (a) DED system using laser together with powder feedstock and (b) using electron beam and wire feedstock and (c) Material Jetting.

Source: For (a) and (b): Taken with permission from Sing et al. (2020); for (c): Taken from Gülcan et al. (2021).

what are their viabilities, what kinds of properties do they offer, and for what processes and applications do we use them. Therefore, material aspects for different 3D printing processes have been elaborated in this section. Based on the state of the materials employed in them, the additive manufacturing (AM) processes are categorized. Based on this classification, there are three

Table 1.1 Advantages and disadvantages of various 3D printing processes

Am Technology	Material Used	Advantages	Limitations	Ref.
VAT Photo-polymerization	UV curable liquid polymer resin	Good finish, high accuracy, relatively fast	Relatively expansive, limited material, post processing time higher	(Halloran, 2016)
Material Jetting	UV curable liquid polymer resin	High Accuracy, Multi material parts	Support material required, limited material	(Sing et al., 2020)
Binder Jetting	Powder of polymer/metal/ceramic and liquid adhesive	Different material and color, faster process	Pot processing needed and hence increase time, Poor part performance	(Gonzalez et al., 2016)
Material Extrusion	Polymer/polymer composite filaments	Easy to operate, less expansive	Poor part performance, limited number of materials, Slow process	(Peng et al., 2018)
Powder Bed Fusion	Powder of metal or polymer	Less cost of production, no support required, Wide range of material	Expansive setup, Low speed, high power, finishing depends upon powder size	(Sutton et al., 2017; Impey et al., 2021)
Sheet Lamination	Sheets of metal or polymer	High speed, low cost, easy handling of materials	Limited materials, delamination during machining	(Derazkola et al., 2020)
Direct Energy Deposition	Polymer and metals	Grain structure control, high strength parts, fast process, repair work	Limited materials, Rough surface finish	(Sing et al., 2020)

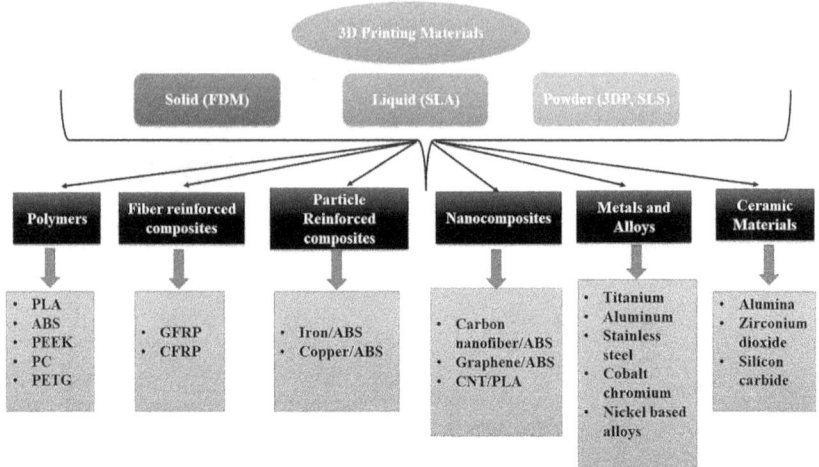

Figure 1.9 Different materials used in additive manufacturing. PLA, Polylactic Acid; ABS, Acrylonitrile Butadiene Styrene; CFRP, Carbon fiber reinforced polymer; CNT, Carbon nanotubes; GFRP, Glass fiber reinforced polymers; PC, Polycarbonate; PEEK, Polyether ether ketone; PETG, Polyethylene Terephthalate Glycol.

Source: Nikzad et al. (2011); Lee et al. (2017); Gu et al. (2012); Van Der Klift et al. (2016); Shofner et al. (2003).

different types of AM processes: liquid, solid and powder. The materials mechanical characteristics should also be suitable and in accordance with service standards. Although the range of materials that can be utilized in various 3D printing procedures is currently limited, work is being done to expand it (Rouf et al., 2022b).

Metals, plastics, composites and ceramics, are the most typical materials utilized in 3D printing. Plastics are widely used in 3D printing. Among plastics thermoplastics are widely used in 3D printing in comparison to thermosetting plastics. Figure 1.9 shows the different types of materials used in 3D printing. The two methods of powder bed fusion and material extrusion both employ thermoplastic materials. Amorphous thermoplastics are employed for material extrusion procedures among these categories because of their melt characteristics. They create a melt that is incredibly viscous and perfect for extrusion (Bourell et al., 2017). Polylactide (PLA) and acrylonitrile butadiene styrene (ABS) are the two most prevalent types of these plastics (ABS). A PC/ABS blend, polycarbonate (PC), and polyetherimide are some further examples of amorphous materials that are utilized in material extrusion (PEI).

Semi-crystalline thermoplastics are employed in powder bed fusion. Polyetheretherketone (PEEK) and polypropylene (PP) are other semi-crystalline polymers utilized in powder bed fusion. Epoxies, acrylics, and acrylates are the most typical thermoset examples. Most of the photo polymers are thermosets. When photopolymers are exposed to light, a process known as "curing" occurs in which the oligomers cross-link and create network polymers, which are thermosets by nature. Oligomers, monomers, and various additives, such as antifoaming agents and antioxidants, are the building blocks of these photopolymers (Sing et al., 2020). These additions improve the photopolymers' characteristics. To enhance the mechanical qualities of resins, toughening compounds are also utilized actively. These hardening agents come in reactive and non-reactive varieties. The elastomeric cores can have a reactive shell in some configurations. polybutadiene, Polysiloxane, and rubber are a few examples of such core materials, whereas substances having epoxy, vinyl ether, hydroxyl, vinyl ester, and acrylate groups are examples of reactive shells. The material jetting technique also employs thermosetting materials. When numerous materials are being deposited, material jetting is advantageous. To do this, various nozzles for various materials are used. The finished product will have distinct qualities from the ingredients that made it up if this procedure is employed to deposit various materials in the same layer. It has been established that the mechanical characteristics of products produced by material jetting exhibit anisotropy while also showing a sizable variation in their tensile and compressive characteristics (Mueller and Shea, 2015).

The two most popular powder-based AM methods for producing metal products, from the perspective of metals, are Powder Bed Fusion (PBF) and Direct Energy Deposition (DED). However, there is an option of using a metallic wire instead of powder in direct energy deposition as well. Metal prints are also created using binder jetting in addition to these two processes. Pure titanium, Ti6Al4V (Niendorf et al., 2013), 316 L stainless (Yadollahi et al., 2015), 17-4PH stainless steel, and 18Ni300 maraging steel (Casalino et al., 2015) are the most widely used 3D printing metals and alloys commercially. Other materials include Co-Cr-Mo, AlSi10Mg (Maskery et al., 2016), and the nickel-based super-alloys Inconel 625 and Inconel 718 (Li et al., 2015). Gold, silver, and platinum are employed as raw materials for selective laser melting, which is then used to print required products (Zito et al., 2014). There are a few explanations for why there aren't many metals suitable for 3D printing. Three-dimensional printed metal parts usually have the problem of high porosity occurring during the printing process as small holes and cavities are formed within the part. Other issues with metal 3D printing are development of residual stresses, cracking and warping, formation of oxides, necessity of post processing and poor surface finish. One such example is the affinity with air in aluminum and aluminum alloys. It causes issues with particle sintering and creates an aluminum oxide layer at the surface. Both Inconel 718 and 18Ni300 maraging steel generate issues in the melt pool by forming stable oxides that climb to the top (Zhang et al., 2013). Silver, copper, aluminum, and gold alloys, which have low absorption and strong thermal conductivity, make it difficult to produce a melt pool. Additionally, the magnitude of residual stresses in metal 3D printing is also a source of concern. High tensile stresses are present at the outside surfaces, whereas zones of compressive stresses are present in the core. Additionally, stress gradients develop within the product and are influenced by its geometry, height, and build direction.

In AM, ceramics are also being used more and more. However, their direct usage in AM presents challenges due to their high melting point and low durability (Niu et al., 2017). Many techniques were used to directly use ceramics in additive manufacturing (AM), but these led to thermally induced cracking. When ceramics are used indirectly in additive manufacturing, the product must be bound together with a binder. All other AM processes, with the exception of direct energy deposition, are utilized in the indirect manufacture of ceramic goods. One of the first techniques used when ceramics were first used in additive manufacturing (AM) was combining ceramic (mostly silicon nitride and alumina) with a stereolithography resin.

A method called freeze-form extrusion fabrication (FEF) uses ceramics to create 3D objects while still being environmentally friendly. It creates the thing layer by layer under the guidance of a computer using aqueous colloidal pastes with minute amounts of organic binder (Leu and Garcia, 2014). One of the main issues with FEF is that large ice crystals form during freezing, causing low final product densities and the development of holes that degrade the product's general qualities. Adopting the Ceramic-on-Demand method, which is done at room temperature and uses radiation to dry the product, is one option to solve this issue. Additionally, it is highly helpful for creating intricate shapes out of ceramic materials (McMillen et al., 2016).

Composites are increasingly being used in 3D printing, and novel composites with unproven qualities are constantly being created. The feedstock material, its qualities, and its uniformity should be taken into account when creating a composite. The composites should be properly bonded together and have strong mechanical characteristics. The three types of composites that are most frequently utilized are metal matrix composites, ceramic matrix composites and polymer matrix composites. In addition to using fiberglass or carbon fiber, fiber reinforced composites are also employed in additive manufacturing (Hofmann et al., 2014). There have been efforts made to find composites that can be produced from waste materials, agricultural byproducts, and items that are both safer for the environment and more cost-effective. One such substance is oil produced from used coffee grounds (Ox-SCG). A single screw FDM filament extruder can produce Ox-SCG and PLA composite filaments. Ox-SCG has been utilized with materials other than PLA in the past. For instance, when researchers (Huang et al., 2018) combined coffee husks with a polyethylene matrix, the total modulus and thermal characteristics increased. Additionally, there have been numerous attempts to use rice husks in 3D printing. In a study, twin screw extrusion was used to combine wood powder and rice husk powder in PLA. Apart from these, many other natural or agricultural by products have been used to fabricate composites for 3D printing such as bamboo (Bodros et al., 2007), coconut, flax, hemp, and wood floor (Deb and Jafferson, 2021).

The development of new materials, or the modification of existing ones, that not only meet the criteria for strength, durability, reliability, economics, and sustainable aspects, but are also compatible with existing 3D printing techniques is required due to the use of smart manufacturing techniques and the need for products with superior properties for advanced engineering and industrial applications. There is still a lot of work to be done in this area.

1.3.1 New product development and 3D printing

The development of new products involves a process from ideation to end of life commonly referred to as product development cycle. With the rise in the competitive global market scenario, it has become imperative to develop products in shorter time durations with reduced cost. The process involves various steps such as ideation, prototyping and testing wherein 3D printing can help. With the introduction of 3D printing technologies, a number of activities involved in a typical product development cycle have become easier, efficient and faster (Wilkinson and Cope, 2015). The stages such as ideation, prototyping involve lot of iterative processes which have become easier by involving 3D printing technologies. The direct development of 3D part from a CAD file helps to achieve designs which are customer centric. Further, the advanced concepts of concurrent engineering such as Design for Assembly, Design for Safety, and Design for Manufacturing can be implemented in a better way while manufacturing by additive manufacturing.

1.3.2 Opportunities, challenges, and application areas

Additive manufacturing has evolved as alternative and efficient manufacturing process particularly for development of parts which include intricacies and complexities in their design. Apart from being material efficient, cost effective the processes are sustainable as well. Also developing optimized geometries with advanced materials has become easy. Further, the cost of tooling and post processing has also considerably reduced. The development of spares for onsite maintenance of remote machinery has become easier. However, there are a lot of challenges which need to be overcome before the large scale implementation and exploitation of these technologies. The challenges include material compatibility of these technologies, cost of raw material, and skilled labor for operation of the machinery in remote locations. Further, the mechanical and physical properties of the parts developed by 3D printing are also not comparable to parts produced by conventional processes. Optimized process parameters and standards needed for producing parts with better mechanical and physical properties are still in developing stage and need more research. Poor interlayer bonding is also a issue which needs to be addressed in most of the AM technologies as the weak interlayer interface leads to crack initiation and hence failure.

The AM parts find applications in diverse fields such as medical, aerospace, solar, automotive, art, jewelry, and construction. The medical applications include dentistry, scaffolds, bone implants, tissue engineering, surgical tools, etc. The application areas can further be expanded by trying new material options and working toward minimization of cost. Involving 3D printing for printing by materials such as smart materials, and nanomaterials can further help to expand the application arena.

REFERENCES

Bajpai, A., A. Baigent, S. Raghav, C. Ó. Brádaigh, V. Koutsos, and N. Radacsi, "4D printing: materials, technologies, and future applications in the biomedical field," *Sustainability*, vol. 12, no. 24, p. 10628, 2020.

Bhatia, A. and A. K. Sehgal, "Additive manufacturing materials, methods and applications: A review," *Mater. Today Proc.*, 2021.

Bodros, E., I. Pillin, N. Montrelay, and C. Baley, "Could biopolymers reinforced by randomly scattered flax fibre be used in structural applications?," *Compos. Sci. Technol.*, vol. 67, no. 3–4, pp. 462–470, 2007.

Bourell, D. *et al.*, "Materials for additive manufacturing," *CIRP Ann.*, vol. 66, no. 2, pp. 659–681, 2017.

Casalino, G., S. L. Campanelli, N. Contuzzi, and A. D. Ludovico, "Experimental investigation and statistical optimisation of the selective laser melting process of a maraging steel," *Opt. & Laser Technol.*, vol. 65, pp. 151–158, 2015.

Chaturvedi, I., A. Jandyal, I. Wazir, A. Raina, and M. I. U. Haq. "Biomimetics and 3D printing-Opportunities for design applications." *Sensors International*, vol. 3, p. 100191, 2022.

Deb, D. and J. M. Jafferson, "Natural fibers reinforced FDM 3D printing filaments," *Mater. Today Proc.*, vol. 46, pp. 1308–1318, 2021.

Derazkola, H. A., F. Khodabakhshi, and A. Simchi, "Evaluation of a polymer-steel laminated sheet composite structure produced by friction stir additive manufacturing (FSAM) technology," *Polym. Test.*, vol. 90, p. 106690, 2020.

Gibson, I. *et al.*, *Additive manufacturing technologies*, vol. 17. Springer, 2021.

Gonzalez, J. A., J. Mireles, Y. Lin, and R. B. Wicker, "Characterization of ceramic components fabricated using binder jetting additive manufacturing technology," *Ceram. Int.*, vol. 42, no. 9, pp. 10559–10564, 2016.

Gorji, N. E. *et al.*, "A new method for assessing the recyclability of powders within Powder Bed Fusion process," *Mater. Charact.*, vol. 161, p. 110167, 2020.

Gu, D. D., W. Meiners, K. Wissenbach, and R. Poprawe, "Laser additive manufacturing of metallic components: materials, processes and mechanisms," *Int. Mater. Rev.*, vol. 57, no. 3, pp. 133–164, 2012.

Gülcan, O., K. Günayd\in, and A. Tamer, "The state of the art of material jetting—A critical review," *Polymers (Basel).*, vol. 13, no. 16, p. 2829, 2021.

Halloran, J. W., "Ceramic stereolithography: additive manufacturing for ceramics by photopolymerization," *Annu. Rev. Mater. Res*, vol. 46, no. 1, pp. 19–40, 2016.

Hirt, L., A. Reiser, R. Spolenak, and T. Zambelli, "Additive manufacturing of metal structures at the micrometer scale," *Adv. Mater.*, vol. 29, no. 17, p. 1604211, 2017.

Hofmann, D. C. *et al.*, "Compositionally graded metals: A new frontier of additive manufacturing," *J. Mater. Res.*, vol. 29, no. 17, pp. 1899–1910, 2014.

Huang, L., B. Mu, X. Yi, S. Li, and Q. Wang, "Sustainable use of coffee husks for reinforcing polyethylene composites," *J. Polym. Environ.*, vol. 26, no. 1, pp. 48–58, 2018.

Impey, S., P. Saxena, and K. Salonitis, "Selective Laser Sintering Induced Residual Stresses: Precision Measurement and Prediction," *J. Manuf. Mater. Process.*, vol. 5, no. 3, p. 101, 2021.

Jandyal, A., I. Chaturvedi, I. Wazir, A. Raina, and M. I. U. Haq, "3D printing–A review of processes, materials and applications in industry 4.0," *Sustainable Operations and Computers*, vol. 3, pp. 33–42, 2022.

Lee, J.-Y., J. An, and C. K. Chua, "Fundamentals and applications of 3D printing for novel materials," *Appl. Mater. Today*, vol. 7, pp. 120–133, 2017.

Leu, M. C. and D. A. Garcia, "Development of freeze-form extrusion fabrication with use of sacrificial material," *J. Manuf. Sci. Eng.*, vol. 136, no. 6, 2014.

Li, S., Q. Wei, Y. Shi, Z. Zhu, and D. Zhang, "Microstructure characteristics of Inconel 625 superalloy manufactured by selective laser melting," *J. Mater. Sci. & Technol.*, vol. 31, no. 9, pp. 946–952, 2015.

Malik, A., M. I. U. Haq, A. Raina, and K. Gupta. "3D printing towards implementing Industry 4.0: sustainability aspects, barriers and challenges," *Ind. Rob. Int J. Rob. Res Appl.*, vol. 49, no. 3, pp. 491–511, 2022.

Maskery, I. *et al.*, "Quantification and characterisation of porosity in selectively laser melted Al–Si10–Mg using X-ray computed tomography," *Mater. Charact.*, vol. 111, pp. 193–204, 2016.

McMillen, D., W. Li, M. C. Leu, G. E. Hilmas, and J. Watts, Designed extrudate for additive manufacturing of zirconium diboride by ceramic on-demand extrusion. International Solid Freeform Fabrication Symposium. 2016. University of Texas at Austin.

Melchels, F. P. W., J. Feijen, and D. W. Grijpma, "A review on stereolithography and its applications in biomedical engineering," *Biomaterials*, vol. 31, no. 24, pp. 6121–6130, 2010.

Mueller, J. and K. Shea, The effect of build orientation on the mechanical properties in inkjet 3D printing. International Solid Freeform Fabrication Symposium. University of Texas at Austin, 2015.

Niendorf, T., S. Leuders, A. Riemer, H. A. Richard, T. Tröster, and D. Schwarze, "Highly anisotropic steel processed by selective laser melting," *Metall. Mater. Trans. B*, vol. 44, no. 4, pp. 794–796, 2013.

Nikzad, M., S. H. Masood, and I. Sbarski, "Thermo-mechanical properties of a highly filled polymeric composites for fused deposition modeling," *Mater. & Des.*, vol. 32, no. 6, pp. 3448–3456, 2011.

Niu, F. Y., D. J. Wu, S. Yan, G. Y. Ma, and B. Zhang, "Process optimization for suppressing cracks in laser engineered net shaping of Al2O3 ceramics," *JOM*, vol. 69, no. 3, pp. 557–562, 2017.

Peng, F., B. D. Vogt, and M. Cakmak, "Complex flow and temperature history during melt extrusion in material extrusion additive manufacturing," *Addit. Manuf.*, vol. 22, pp. 197–206, 2018.

Rouf, S., A. Raina, M. I. U. Haq, N. Naveed, S. Jeganmohan, and A. F. Kichloo, "3D printed parts and mechanical properties: influencing parameters, sustainability aspects, global market scenario, challenges and applications," *Advanced Industrial and Engineering Polymer Research*, vol. 5, no. 3, pp. 143–158, 2022a.

Rouf, S. *et al.*, "Additive manufacturing technologies: industrial and medical applications," *Sustain. Oper. Comput.*, vol. 3, no. 258–274, 2022b.

Saxena, P., G. Bissacco, K. Æ. Meinert, A. H. Danielak, M. M. Ribó, and D. B. Pedersen, "Soft tooling process chain for the manufacturing of micro-functional features on molds used for molding of paper bottles," *J. Manuf. Process.*, vol. 54, pp. 129–137, 2020.

Saxena, P., E. Pagone, K. Salonitis, and M. R. Jolly, "Sustainability metrics for rapid manufacturing of the sand casting moulds: A multi-criteria decision-making algorithm-based approach," *J. Clean. Prod.*, vol. 311, p. 127506, 2021.

Shim, D.-S., G.-Y. Baek, J.-S. Seo, G.-Y. Shin, K.-P. Kim, and K.-Y. Lee, "Effect of layer thickness setting on deposition characteristics in direct energy deposition (DED) process," *Opt. & Laser Technol.*, vol. 86, pp. 69–78, 2016.

Shofner, M. L., K. Lozano, F. J. Rodríguez-Macías, and E. V. Barrera, "Nanofiber-reinforced polymers prepared by fused deposition modeling," *J. Appl. Polym. Sci.*, vol. 89, no. 11, pp. 3081–3090, 2003.

Sing, S. L., C. F. Tey, J. H. K. Tan, S. Huang, and W. Y. Yeong, "3D printing of metals in rapid prototyping of biomaterials: Techniques in additive manufacturing," edited by Roger Narayan in *Rapid prototyping of biomaterials*, Elsevier, 2020, pp. 17–40.

Sivarupan, T. *et al.*, "A review on the progress and challenges of binder jet 3D printing of sand moulds for advanced casting," *Addit. Manuf.*, vol. 40, p. 101889, 2021.

Sutton, A. T., C. S. Kriewall, M. C. Leu, and J. W. Newkirk, "Powder characterisation techniques and effects of powder characteristics on part properties in powder-bed fusion processes," *Virtual Phys. Prototyp.*, vol. 12, no. 1, pp. 3–29, 2017.

Vaezi, M., H. Seitz, and S. Yang, "A review on 3D micro-additive manufacturing technologies," *Int. J. Adv. Manuf. Technol.*, vol. 67, no. 5, pp. 1721–1754, 2013.

Van Der Klift, F. *et al.*, "3D printing of continuous carbon fibre reinforced thermoplastic (CFRTP) tensile test specimens," *Open J. Compos. Mater.*, vol. 6, no. 01, p. 18, 2016.

Wilkinson, S., and N. Cope, "3D printing and sustainable product development," edited by Mohammad Dastbaz, Colin Pattinson and Babak Akhgar in *Green information technology*, pp. 161–183. Morgan Kaufmann, 2015.

Yadollahi, A., N. Shamsaei, S. M. Thompson, A. Elwany, L. Bian, and M. Mahmoudi, Fatigue behavior of selective laser melted 17-4 PH stainless steel. International Solid Freeform Fabrication Symposium. University of Texas at Austin, 2015.

Zhang, Y. N., X. Cao, P. Wanjara, and M. Medraj, "Oxide films in laser additive manufactured Inconel 718," *Acta Mater.*, vol. 61, no. 17, pp. 6562–6576, 2013.

Zito, D. *et al.*, "Optimization of SLM technology main parameters in the production of gold and platinum jewelry," in *The Santa Fe Symposium on Jewelry Manufacturing Technology 2014*, vol. 47, no. 13, pp. 439–470, 2014.

Chapter 2

How 3D Printing Achieves Sustainability

Haishang Wu, Nida Naveed, and Hamid Mehrabi

CONTENTS

2.1 INTRODUCTION

Sustainability is becoming critical in this decade. The Brundtland Report (1987), published by the World Commission on Environmental Sustainability and Development, is recognized as a pioneering document that brought attention to sustainability concerns. The report tackles various pressing issues that necessitate immediate action, including energy efficiency, the global impact of industrial growth and sustainable development. Toward this, sustainability can be analyzed from multiple dimensions, such as environment, society and economy. In this chapter, plastics materials are applied, as an example of the recycling and manufacturing process, to illustrate how additive manufacturing (three dimensional printing - 3DP) can achieve sustainability (Figure 2.1).

DOI: 10.1201/9781003306238-2

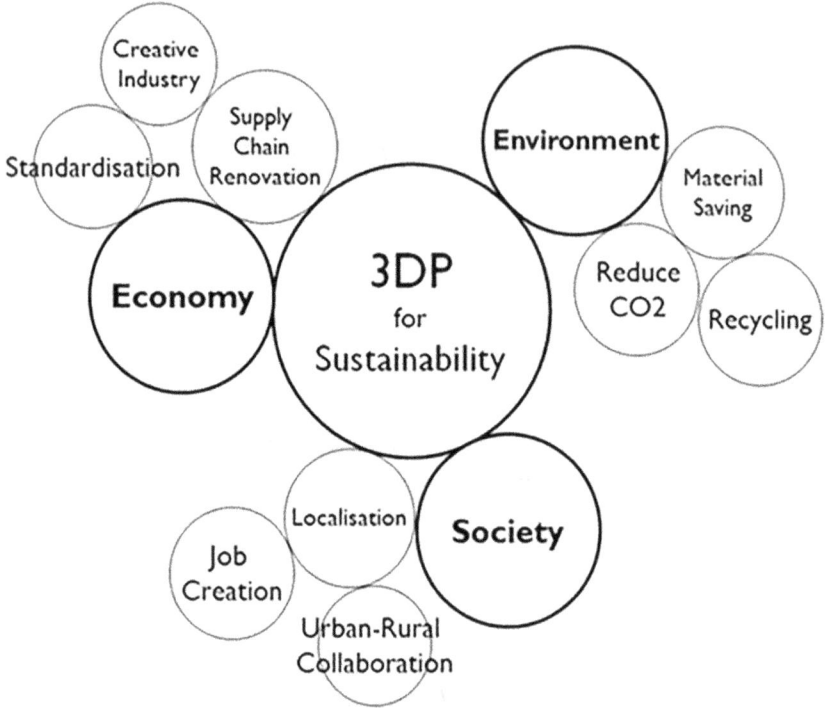

Figure 2.1 The roles of additive manufacturing (3DP) in promoting sustainability of environment, society and economy.

The role of 3DP toward sustainability in terms of environment, society and economy is discussed. Three-dimensional printing and its impacts against materials recycling and manufacturing, products applications, the advantages and disadvantages, AM's promotion of humanism, as well as how 3DP affects transportation, localization, and supply chain are discussed in the following 12 sub-sections. Moreover, issues such as rural development and transportation mitigation, strategic planning and standardization are addressed.

2.2 RECYCLING

Materials recycling can reduce the materials wasteat the end of life (EOL) to those countries emphasising the materials recycling and recycling technologies, while producing tremendous socioeconomic values (Peeter et al., 2017). Methods for how to increase recycling rate has been a critical mission, as the material recycling rates are still low in developed countries, while the rates of materials recycling are still close to zero in those developing countries (d'Ambrières, 2019). For this reason, improvement of materials recycling rate and recycling efficiency is one of key elements to the sustainability of manufacturing industry. Furthermore, reduction of materials waste through the recycling process can be advantageous to the 3DP technologies.

However, based on the statistics of materials waste in 2015, the materials recycling accounted for less than 20% globally, while disposal accounted for 55% and incineration accounted for over 25%. In order to increase the materials recycling rate, the ecologists encourage enterprises and various sectors to form synergies to tackle this issue (El-Haggar and Salah, 2007). In this study, the objective is to achieve 100% materials recycling with 0% landfill which can be challenge. However, this is feasible if the materials recycling cost can be minimized and efficiency can be improved. Eventually, consumption of energy and emissions of CO_2 can be eliminated, and the yield (%) of materials can be improved (Wiesmeth, 2020).

Fast development of AM has built a foundation ofeconomical sustainability of distributed-recycling-AM (DRAM) (Little et al., 2020). Three-dimensional printing can print parts of a product by computer-aided design (CAD) by using standard tessellation language (STL) or stereo-lithography files holding the data of those sliced shape or triangles of the parts that can fully utilize the recycled materials in various applications (Wong and Hernandez, 2012).

2.3 PRODUCTS APPLICATIONS

Three-dimensional printing technology emerged in the 1980s. In these few decades, it has grown exponentially in the fields of technology, market and applications. Three-dimensional printing is expected to maintain a steady growth of 30.2% annually, reaching $34.8 billion by 2026 (Markets and Markets, 2021).

The ASTM Standard categorizes AM into seven types of technology, and they comprise: Extrusion of Material, Sheet-lamination, Vat-photo-polymerization, Binding-jetting, Powder-bed-fusion, Directed-energy deposition and Material-jetting (ASTM International, 2013). Deployments of the AM technologies depend of the applications and conditions of the manufacturing process, and the usages and selection of technology can be critical to the 3DP or AM processes (Santander et al., 2020; Özkan et al., 2015).

Three-dimensional printing manufacturing is carried out via different methods. The materials can besolid, liquid or powder, and different types of materials have their respective advantages. Main componentsfor AM machine are: the frame, printer head and movement mechanics, build platform, stepper motors, electronics, firmware, software and filaments as support substances (3D Insider, 2020).

Consumer behaviors in reducing materials usages and recycling habits can also be important for the recycling waste management. Recycled materials wastes were tracked and analyzed, to produce the statistics of recycling rate. As it is indicated in the global statistics in 2015; over 95% of the 146 million tons of primary materials was disposed which threats the sustainability of the manufacturing industry from the viewpoints of the sustainability of environment, economies and society (Dengler, 2017).

Within the plastics industry, high-density-poly-ethylene (HDPE), polypropylene(PP), polyethylene-terephthalate (PET), and low-density-polyethylene (LDPE) accounted for 62% of global plastic demand (PlasticsEurope Market Research Group, 2015). Among these, PET and PP contribute 45% of plastic products, while 60% of both types are deployed in the applications of packaging. Such type of application is controllable if the consumers have sufficient awareness of sustainability and willing to corporate by following the regulations or instructions. From the viewpoints of manufacturers, types of the recycled plastics vary which mainly consist of flake, pellets or materials that do not need extensive processing. If the manufacturers can widely cover all different types of recycled plastics, and reduce the cost and improve the process efficiency, then the plastics recycling rate can be significantly improved.

2.4 ADVANTAGES OF 3DP

AM can be one of the best methods in plastics industry, as AM covers a broad range of functions, shape, and complexity. Example is, AM deployment in medical instruments fully meets the requirements of customization and personalization that may not be easily replaced by other technologies (EOS, 2021). The other example is the AM applications in the automotive industry that requires a frequent replacement of parts or components of different brands or different structures of the vehicle that only AM can print the parts based on the demands (Zahnd, 2018).

The AM's characteristics in design flexibility make prototyping easy, and AM can manufacture the complex productions of different shapes and various types of materials without difficulty. Furthermore, AM can be cost-effective and time saving in those suitable applications (Hendrixson, 2016; Kim et al., 2021). A robust integration across plastics recycling and process of AM can be even an efficient way in achieving sustainability and forming a closed-loop recycling which is important to sustainable manufacturing. Closed-loop recycling enables the entire designing process, materials, prototyping, manufacturing and products to be connected together in a sustainable loop (Hendrixson, 2021).

In terms of the products, as 3DP is able to change some particular parts without impacting the whole item, it is being increasingly used in lightweight products that require high customization, energy-saving and personalization, and also the products that require frequent replacement of their parts or components (Huang et al., 2016).

AM applications are estimated to hit $23.33 billion in 2026, and among all products, AM is expected to achieve high performance and light-weight requirements in aerospace industry, as well as those high demands in medical industry based on its unique characteristics in personalization and customization. The advantages of AM and successful stories in the typical applications are listed as followed:

- **Transportation:** Reduction in transportation distance is the key advantage of AM. The localization of AM can reduce cost, supply chain requirements (Garmulewicz, 2016), transportation distance, energy consumption, and CO_2 emission.
- **Medicare industry:** High personalization and customization are the basic requirements of medical instruments. Applications in medical industry are versatile, from eyeglasses, hearing aids, orthodontics, to the tissue repair and organ transplantation. An emerging technology of Medicare has been developed by the State University of New York. The method "FLOAT" is able to print human body parts in a few minutes (Anandakrishnan et al., 2021). This development enables AM capability with high potential to save millions of lives, by allowing rapid organ transplantation.
- **Architecture industry:** AM in architecture industry is promising in construction. AM advantages cover; environmental friendliness, low cost, reduced injury, and time saving (Hager et al., 2016). AM enables users flexibility in design and adjustment based on their own habitat in simple in-situ construction. These advantages are particularly important to the developing societies and urgent conditions such as aftermath of disasters.
- **Automotive industry:** One automotive can be composed by hundreds or thousands of parts, and capabilities in prototyping and printing spare parts are critical. Because 3DP does not need a final assembly process, 3DP or AM can be flexible in fabricating spare parts based on demands, and cut the manufacturing life cycle, and these are the unique characteristics of AM or 3DP. For instance, AM Company, Stratasys, improves the efficiency for the jigs and fixtures department of BMW, which has led the car manufacturers deciding to use AM as the primary method replacing the conventional manufacturing (CM) (The 3D Printing Solutions Company, 2014).

Based on the analysis work of Frost and Sullivan (Frost & Sullivan Global Research Team, 2016), 3DP will continue its steady growth across different regions. Among all regions, Asia ranks top at 55% yearly growth based on the trend and estimation. The scale of 3DP isgrowing in all continents however, even though 3DP is not a labour-intensive technology, 3DP relies on home-based businesses (HBM) to enable its capability in mass production, and to meet the markets' demands. For this reason, localization of

3DP or AM, and a seamless integration to plastics recycling is crucial to fully utilize 3DP's advantages.

2.5 CHALLENGES OF 3DP OR AM

Compared to primary plastics, reuse of recycled plastics is critical to sustainability as plastics recycling saves materials, reduces energy consumption and the cost as well. However, plastics recycling needs an improvement in the recycling process to sort the plastics waste, and to eliminate the quality degradation as in the current situation, a quality degradation at around 10% is common in plastic recycling process (Merrild et al., 2012).

In addition, 3DP or AM is facing a challenge in speed and scale that can limit 3DP to those products of smaller quantity with higher complexity (Lee et al., 2017). Fully utilizing the HBM, and innovation of technologies of 3DP may solve the speed and scale issues (GE Additive, 2020). For instance, industry is developing new AM technologies that deploying up to one million of diode lasers to increase the scale and to enhance the printing speed (AMFG, 2020). In addition, 3DP is expected to fully utilize multi-entities through a collaborative HBM community that may minimize the challenge of speed and scale.

2.6 3DP FOR HUMANISM

Three-dimensional printing is widely recognized and utilized mainly for business and aesthetic usages. However, the technology also has huge potentials to be developed for humanitarian uses. Firstly, AM allows all users to design and customize personal items, especially in times of emergency. During the early stage of the COVID-19 pandemic, there was severe shortage of facial masks. Two students in Japan collaborated to print self-designed mask "PITATT" and provided open-source code (Iju and Hattori, 2021). As changing product shapes only requires changing the software codes therefore, quality improvements can happen with minimum efforts, particularlyto those items of common usages with huge number of users.

AM assists development and job creation especially in rural areas by means of HBM. Three-dimensional printing machines for entry levels usually cost only a few hundred US dollars (Carneiro et al., 2020). Some websites such as "Repetier.com" provide free CAD software of 3DP which can be saved in a repository and can be accessed world-widely through the cloud technologies. The reusable files and easy manipulation of the software codes are stored in a standardized format, and "printing on demands" can be one of the most significant advantages of 3DP that save time, increasing efficiency, reduce cost and contribute to sustainability (Michelle, 2018). Basic AM design does not require high level of skills and capital for production, and most of the people can establish home business and become producers. For those local cultures, art communities and households, AM can empower designers to improvise with unique creations or folk art.

2.7 TRANSPORTATION

Transportation has been identified as a significant factor that impacts sustainable manufacturing. A full utilization of recycled plastics replies on the efficiency of plastics recycling through an optimization process. In this optimization process, transportation distance can be optimized under a precise calculation of plastics recycling facilities that minimize unnecessary transportation, and significantly reduce energy consumption and CO_2 emissions. Supported by the simulation technique, the optimized transportation of plastics recycling process can reduce the cost, CO_2 emission and energy consumption (Wu et al, 2022).

Localization has been one of the most critical characteristics of AM or 3DP that contribute to sustainability if the plastics recycling and manufacturing are fully integrated. Based on this foundation, a robust integration of plastics recycling and AM can reduce transportation distance and the consumption of energy however; this critical factor has been missing in the existing literature. Most of the literature consider that AM may not easily achieve process standardization through the integration (Peng et al., 2020) which may not reflect reality instead; they may cause a slowdown of AM advancement.

From the viewpoint of environmental sustainability, transportation can be a significant factor and localization may ease the environmental threat caused by plastics waste transportation as supply chain can be eliminated in AM and all parts do not need to be assembled in a final plant (Garmulewicz, 2016). For this reason, this study utilizes the AM's advantages to eliminate supply chain to reduce transportation and to improve plastics recycling and AMprocess to achievebetter sustainability in environment.

Minimizing supply chain engagement in 3DP is feasible, but a full elimination is not realistic. As it is indicated by Attaran (2017); the five areas that transform supply chain process; decentralization, sustainable manufacturing, competition, customization, and manufacturing on-demands. This improvement may strengthen AM's capability in reducing transportation before a full elimination of supply chain.

2.8 LOCALIZATION

Traceability, scaling production, quality and speed are the barriers to the current3DP technology need improvement. Localization is the critical factor to AM that support traceability, liability and legality as all the transactions are stored in the cloud environment. A sharable platform and community for

AM helps the traceability in the decentralization environment, and transparency of a common platform enhances the traceability, which can be the first step of localization.

From the existing literatures, AM's local recycling and manufacturing will become a trend with the advancement of technologies of AM (Kleera and Pillerb, 2019). More literatures further indicate that as assembly and tooling are not mandatory steps to AM, AM's dependency on logistics and supply chain can be eliminated. AM tips the balance between manufacturing and recycling, between outsourcing and insourcing of supply chain, and between recycling of plastics and AM process that can advantageous (Mourdoukoutas, 2015).

Furthermore, 3DP's easy-entry to HBM supports localization and drives AM's advancement into a more decentralization in a distributed manufacturing environment (Inimake, 2021). In this transformation, HBM will become a significant workforce of 3DP that supports AM in resolving scale and speed issues and meanwhile, solve society issues as tremendous job opportunities can be created in rural areas through the development of AM community or HBM. Because setting up of 3DP machine only requires a few hundreds US dollars to start HBM, it will be easier for the new residents to establish their AM career if the authorities can support the infrastructure, education, and regulations in the relocation from urban to rural areas. Under a solid foundation of AM community and HBM, localization minimizes supply chain reliance and transportation distance.

Ben-Ner predicted that localization will be the trend in the future and that extensive supply chain will be reduced or replaced by alternative solutions (Ben-Ner et al., 2017). Meanwhile, local network will replace long supply chain and support collaboration in the sustainable supply chain. Several literatures suggested AM local supply is feasible and cost-effective, but technologies and standards can be conditions and the areas of challenges. In addition, transportation can be a crucial factor of supply chain from both economies and environment perspectives (Garmulewicz, 2016, Attaran, 2017, Akbari and Ha, 2020).

Distributed and decentralized manufacturing are the foundation of localization. In a digital world, global collaboration and local manufacturing will establish flexibility in plastics industry and bring in positive impacts on sustainable development (Attaran, 2017). Local manufacturing will become a trend with the advancement of AM technologies (Kleera and Pillerb, 2019), and local manufacturing in a distributed environment (Kleera and Pillerb, 2019) can be the step stones of HBM and AM community in this advancement.

2.9 SUPPLY CHAIN

Supply chain and logistics can cause enormous energy consumption and CO_2 emission if they are not well-designed. Three-dimensional printing through plastics recycling can eliminate supply chain and reduce energy consumption. With the advancement of AM technologies, decentralization of AM can save the cost, reduce risk and lead time, while increasing operational autonomy (Khajavi et al., 2013).

Reduction of reliance on supply chain gives us a better future in sustainability and manufacturers of plastics shall establish collaboration with the suppliers in this transformation (AiChin et al., 2015). Products produced by AM are a further step close to those consumers and less demand on packaging, transportation and materials for shipping can save cost and protect our environment that contribute sustainability (Kubáč and Kodym, 2017).

AM prevents constraints from traditional supply chain, particularly for those low volume, and customer-specific items. Its flexibility and agile adaptation to demands engenders many benefits unwitnessed by traditional supply chain, such as customised production, localized distribution and manufacture, short transportation distance and lead time, and low footprint of carbon (Kubáč and Kodym, 2017). In taking sustainability concept into supply chain management, green supply chain has become a trend recently (Mohtashami et al., 2020) and reduction in transportation distance can be a key area of green supply chain that consume less energy and produce less CO_2 (Nikolaoua et al., 2013) that support environmental sustainability.

From the perspectives of economies, cost in transportation can be expensive, and the final price is often several times of the original value. The transportation requires natural and economic resources, resulting in fuel usage, air pollution and extra cost (Mckinnon and Yongli, 2006).

In addition to the technologies and process, there is a mutual dependency between plastics recycling rate and cost saving which requires collaboration and standards to enable their interferences. In addition, the concept of "fabricating parts on demands" can reduce supply chain reliance and can reduce plastic waste as well (Ribeiro et al., 2020). However, these are not found in the related literatures, which need further investigation.

2.10 RURAL DEVELOPMENT AND TRANSPORTATION MITIGATION

Many literatures indicate that the AM localization can be established through the hub-like networks of local communities. In this hub, HBM plays significant role in providing workforce of AM through the relocating of the overcrowded population in urban areas to form AM community in rural local areas, which can be also a win-win situation for both urban and rural areas. As indicated in the World Bank, overcrowing situation in big cities will hit 70% of global population by 2050 (World Bank population density, 2018). The rapid growth of population affects the rural development, and degrades environmental sustainability (Ray, 2011).

From the perspective of AM workforce, overcrowding in major cities causes abnormal growth of energy consumption and degrades of ecosystem. To mitigate the issues, the U.S. has developed several programs for AM, to fabricate parts and create opportunities of AM jobs in rural development (Legg, 2021). The programs also provide training in manufacturingto help the residentsto establish their careers within the local area (Jones et al., 2021).

2.11 STRATEGIC PLANNING

To promote AM to an industry mainstream, a realization plan needs to be in place. Currently, AM needs some improvements in its technologies and processes, to make it efficient. In addition, the cost pattern, and workforce of AM need to be planned first, followed by an evaluation, particularly the applicability of the standards and methods, to tackle with AM challenges, and to fully utilize AM's advantages to convince stakeholders to invest AM in those areas need further development. Through a robust integration between plastics recycling and AM process in rural areas, the win-win results can be expected, and sustainability can be achieved.

Similar to other industry, plastics recycling and manufacturing have close linkage to the developmentin rural areas. This implies, rural development requires stable residency from the local people, to devote their effort to AM community or HBM. However, abnormal migration has caused significant impact that affect rural development. Migration from rural to urban in an abnormal situation can ruin the plan of rural development and simultaneously affect the sustainability in the urban areas as well. Social-economists alerts that; the povertycan be the tactical issue and one of the major reasons cause abnormal migration. In this over-urbanization situation, the surplus workforce in urban areas can fasten situation of unemployment in the lower-income level and potentially hastens an unbalanced ecosystem, living quality degradation, pandemic crisis, crime records and social issues that affect sustainability (Gebrea and Gebremedhinb, 2019).

Settlements Program from the UN Humanity also alerted that abnormal rural to urban migration can affect sustainability of environment, economies and society, particularly the ecosystem (Gebrea and Gebremedhinb, 2019). In order to mitigate this issue, job opportunity in rural areas, rural development and livelihood of rural residents need to be handled carefully to make sure the rural habitants have same access and equal quality of infrastructureto improve their livelihood toward a resilient society.

Overcrowdedness in major cities can degrade living quality and affect sustainability. Relocating those congested population, from urban to rural areas, will not only lessen this degradation but also provide a robust workforce to the rural development of AM or HBM. Job opportunities in rural areas can attract tremendous AM workers and HBM to establish their career in rural development, and support AM localization toward sustainability. Consequently, the logistics and supply chain can be eliminated through AM localization, transportation distance can be minimized, and consumption of energy and emission of CO_2 can be reduced.

AM localization will become a trend with the advancement of technologies (Kleera and Pillerb, 2019). Through a well-planned localization, the demographical balancing will support AM to reduce supply chain and increase workforce (Akbari and Ha, 2020; Arora et al., 2021).

In order to establish a concrete foundation of rural development supporting AM or HBM, this research proposes a model of strategic control to guide the planning and metric usages to fully utilize AM's advantages. Furthermore, the model enables the sourcing capability of workforce to assist AM's mass production and manufacturing on demands by relocating the population from congested cities to the rural areas (Arora et al., 2021). The model not only creates tremendous job opportunities but also supports social sustainability, solving many social issues. Through this transformation, the new residents in rural areas have opportunities to enjoy their equal opportunity of livelihood same as those people in urban areas (Kjaerheim, 2005).

In this model, the population balancing plays a critical role in strategic planning because it realizes AM localization. With the advancement of AM technology, a huge portion of the nation's land can be fully utilized through the rural development in AM and HBM. This study also discovers some use cases in the recent AM development through international collaboration. For instance, the huge AM market has promoted some great opportunities in India demanding huge workforce in the suburb or rural areas. However, in order to prevent the risk and build a concrete foundation in sustainability, the strategic control needs to be in place before the mass production.

In terms of control metrics, the rural population ratio and crowdedness index are the important indicators in setting up regulations of demographic population to establish a virtuous cycle in sustainability. Strategic control model provides guidelines through a human-centric approach, and fills the gap in AM technologies or process, and to envision the correlations of different aspects in sustainability.

2.12 STANDARDS

Standards mean uniform formats or specifications in technologies, processes or the methods that can be commonly recognized by the industry in the daily operation. The standards need public agreement as promulgated in those recognized organizations of standards (Clark, 2017). There are many areas in AM need standardization such as technologies, materials, tests, printing parameters or product specifications. Among these, testing standards can be crucial which can be the foundation of AM technologies that making the products reliable and reproducible (Dizona, 2018).

The responsibility of AM standards shall not be limited to one person or one role. AM production requires multi-entities to consolidate different viewpoints into one vision that is sharable across common practices. Currently, as lack of communication and consensus in the plastics industry and large enterprises, AM standards are still a challenge.

In order to solve the issue, this research proposes collaboration and communication across different entities, and suggests a broad scope communications; such as large enterprises, authorities of government, academia, designers, consumers and manufacturers are recommended to develop the forums of discussion. In addition, this research suggests a construction of common repository and platform to share the design concept, results of product tracking and software, to establish a uniform format for common usages. This supports the standardization of AM to become industry mainstream.

In 2009, F42 committee was formed by ASTM international to develop global standards of AM technologies and materials that covered a variety of applications, materials and technologies. In 2011, ISO/TC committee was formed by ISO, and both ISO and ASTM jointly agree to establish global standards through this join development to make the standards more transparent and ensure their full coverage.

Initially, the standards covered products quality, safety, machinery, and the working environment of AM. On the other hand, the Standard Development Organisations (SDOs) developed specific standards in the AM industry domain (Gumpinger et al., 2021) that may supplement some additional requirements of AM in the particular sectors. These standards comprise materials, design, applications, test methods, and terminologies to make sure the quality of the products and procedure satisfy AM's requirements (ASTM International, 2021).

In this consolidation, as evaluation takes time, and verification and validation require long process to ensure the accuracy in terms of the fabrication process, surface finish, data format, and processing time, the issuance of certificate may not be efficient to cope with those urgent cases. Furthermore, the intellectual properties, legal process and mechanism of monitoring are all the limitations that need to be covered in the global standardization.

Through the consolidation of global standardization, STL files, CAE, CAD and the types of materials and their properties are well defined through standardization process. In this process, as mechanical and materials properties need rigorous testing particularly in some specific usages, the software as listed above need to be standardized in the global practices, hence the designers can use the software directly, to print the products in various applications and do not need a conversion during the printing (Wong and Hernandez, 2012).

2.13 CONCLUSION

In this chapter, the role of 3DP toward sustainability is discussed. Three-dimensional printing technology contributes to environment sustainability mainly through the waste recycling, materials saving, and CO_2 reduction. To the 3DP industry, the higher recycling rate means less material consumption. To the individual, 3DP's capability in the engagement of local businesses can eliminate supply chain, minimize transportation and contribute to reduction of GHG emissions and energy consumption. Three-dimensional printing also supports the households to utilize the recycled materials in the spare parts and amend components in a meticulous and agile manner, and effectively reduce wastes.

From social sustainability perspectives; localization, job creation and collaboration are the potential areas that 3DP can fully utilize its natural characteristics. Three-dimensional printing can be one of the top humanist technologies, empowering individual consumers, households and small local businesses based on demands. It is capable of fulfilling the needs for niche and customized items, such as those medical usages that require high degree of personalization. In those cases of emergency, 3DP also serves as a swift aid for disaster aftermaths, to supply the demanding items locally; or to meet the sudden surge of huge demands such as demanding from the pandemic crisis. Furthermore, 3DP alleviates the excessive dependency of individuals toward the market, transforming consumers into prosumers.

As to the economic sustainability, supply chain renovation, creative industry and standardization are the discussion topics. Three-dimensional printing can effectively reduce costs, lead time and tooling and machinery efforts, which greatly benefits industries. Meanwhile, individuals and HBM are capable of manufacturingparts or the whole products at a lower cost and higher flexibility. Under the support of localization and HBM, the current supply chains can be simplified, shortened or even prevented.

In addition to its miscellaneous applications, creative industry is one of the most favorable fields, which can be an interesting area of 3DP to accomplish within a short time. Creative industries, such as crafts, wearable accessories, ornaments, toys and utensils, are some of the most direct visual representations of the cultures and the values of a community. Empowered with beginners-friendly technology like 3DP, local communities can visualize and materialize their concepts in the simplest and most effective ways.

Finally, standardization of 3DP technology is discussed and solutions of safety are suggested. Among all organizations of standards, ISO and ASTM are the most influential providers of international and national standards that contribute global standardization. A humanistic approach of how 3DP achieves sustainability is the core of discussion. Governmental authorities shall take responsibility and play a key role to assist the safe usages of 3DP for local communities.

REFERENCES

3D Insider, (2020), "The 9 different types of 3D printers", accessible at: https://3dinsider.com

AiChin, T., (2015), "Green supply chain management, environmental collaboration and sustainability performance", Procedia CIRP, vol. 26, pp. 695–699.

Akbari, M., Ha, N., (2020), "Impact of additive manufacturing on the Vietnamese transportation industry: An exploratory study", The-Asian-Journal-of-Shipping-and-Logistics, vol. 36, Issue 2, pp. 78–88.

AMFG, (2020), "10 of the biggest challenges in scaling additive manufacturing for production in 2020", https://amfg.ai/2019/10/08/10-of-the-biggest-challenges-in-scaling-additive-manufacturing-for-production-expert-roundup/

Anandakrishnan, N. et al. (2021), "Fast stereolithography printing of large-scale biocompatible hydrogel models", Advanced Healthcare Materials, pp. 2002103, DOI: 10.1002/adhm.202002103

Arora, P., Arora, R., Haleem, A., Kumar, H., (2021), "Application of additive manufacturing in challenges posed by COVID-19", Materials Today, vol. 38, pp. 466–468.

ASTM International, (2013), "ASTM F2792-12a. Standard terminology for additive manufacturing technologies", accessible at: chrome-extension://efaidnbmnnnibpcajpcglclefindmkaj

ASTM International, (2021), "Additive manufacturing technology standards", can be accessed at: https://www.astm.org

Attaran, M., (2017), "The rise of 3-D printing: The advantages of additive manufacturing over traditional manufacturing", Business Horizons, vol. 60, pp. 677–688, accessible at: 10.1016/j.bushor.2017.05.011

Ben-Ner, A., Siemsen, E., (2017), "Decentralization and localization of production: The organizational and economic consequences of additive manufacturing (3D printing)", California Management Review, vol. 59, pp. 5–23.

Carneiro, H., et al. (2020), "Additive manufacturing assisted investment casting: A low-cost method to fabricate periodic metallic cellular lattices", Additive Manufacturing, vol. 33, p. 101085.

Clark, J., (2017), "3D-printing opportunity for standards: Additive manufacturing measures up", Delloitte Insight, accessible at: https://www2.deloitte.com/us/en/insights/focus/3d-opportunity/additive-manufacturing-standards-for-3d-printed-products.html

d'Ambrières, W., (2019), "Plastics recycling worldwide: Current overview and desirable changes", The Journal of Field Actions, Special Issue 19, 2019.

Dengler, R., (2017), "Humans have made 8.3 billion tons of plastic. Where does it all go?", PBS News Hours, can be accessed at: https://www.pbs.org

Dizona, J., (2018), "Mechanical characterization of 3D-printed polymers", Additive Manufacturing, vol. 20, pp. 44–67.

El-Haggar, S.M., Salah M., (2007), "Sustainable industrial design and waste management", Elsevier Academic Press.

EOS, (2021), "Complex-geometries-are-possible-with-3D-printing)", EOS, https://www.eos.info/en

Frost Research Team, (2016), "Global additive manufacturing market, Forecast to 2025", https://namic.sg/wp-content/uploads/2018/04/global-additive-manufacturing-market_1.pdf

Garmulewicz, A., (2016), "Redistributing material supply chains for 3D-printing", 3DP-RDM Dissemination and Scoping Workshops, https://www.ifm.eng.cam. ac.uk

GE Additive, (2020), "Speed and scale in the spinal sector", http:// additivemanufacturing.com

Gebrea, T., Gebremedhinb, B. (2019). "The-mutual-benefits-of-promoting-rural-urban interdependence-through linked-ecosystem-services", Global-Ecology-and-Conservationv, vol. 20, p. e00707.

Gumpinger, J., et al. (2021), "Recent progress on global standardization in additive manufacturing materials and technologies", Fundamentals-of-Laser-Powder-Bed-Fusion-of-Metals, Elsevier, 2021, pp. 563–582.

Hager, A., Golonka, A., Putanowicz, R., (2016), "3D printing of buildings and building components as the future of sustainable construction", Procedia-Engineering, vol. 151, pp. 292–299.

Hendrixson, S., (2016), "Agility through 3D printing", Additive Manufacturing, https://www.additivemanufacturing.media

Hendrixson, S., (2021), "Additive manufacturing will aid and accelerate the circular economy", Additive Manufacturing, https://www.additivemanufacturing.media

Huang, R. et al. (2016), "Energy and emissions saving potential of additive manufacturing: the case of lightweight aircraft components", Journal-of-Cleaner-Production, vol. 135, Issue 1, pp. 1559–1570.

Iju, C., Hattori, T., (2021), PITATT cool 3D print mask, accessible at: https:// pitatt3dprintmask.wixsite.com/website

Inimake, (2021), "3D-printing & additive manufacturing news", Additive News, https://additivenews.com

Jones, F., Mardis, M., Prajapati, P., Kowligi, P., (2021), "Facilitating advanced manufacturing technicians' readiness in the rural economy: A competency-based deductive approach", ASEE-Conference-2021 July.

Khajavi, S., Partanen, J., HolmstrÖm, J., (2013), "Additive manufacturing in the spare parts supply chain", Computers-in-Industry, doi: http://dx.doi.org

Kim, H. et al. (2021), "Additive manufacturing of high-performance carbon-composites: An integrated multi-axis pressure and temperature monitoring sensor", Composites Part B: Engineering, vol. 222, 1 October 2021, p. 109079.

Kjaerheim, G. (2005), "Cleaner production and sustainability", Journal-of-Cleaner-Production, vol. 13, Issue 4, pp. 329–339.

Kleera, R., Pillerb, F., (2019), "Local manufacturing and structural shifts in competition: Market dynamics of additive manufacturing", International-Journal-of-Production-Economics, vol. 216, pp. 23–34.

Kubáč, L., Kodym, O., (2017), "The impact of 3D printing technology on supply chain", MATEC-Web-Conf. 18th-International-Scientific-Conference – LOGI 2017, vol. 134.

Kubáč, L., Kodym, O., (2017). "The impact of 3D printing technology on supply chain", MATEC Web of Conferences, vol. 134, p. 00027. 10.1051/matecconf/201713400027

Lee, J., Chee, J., Chua, K., (2017), "Fundamentals and applications of 3D printing for novel materials", Applied-Materials-Today, vol. 7, pp. 120–133.

Legg, H., (2021), "Rural development project uses 3D printing in fight against COVID-19 spread", U.S. Department of Agriculture, can be accessed at: https:// www.usda.gov

Little, H., et al. (2020), "Towards distributed recycling with additive manufacturing of PET flake feedstocks", Materials 2020, vol. 13, p. 4273.

Markets and Markets, (2021), "3D printing market with COVID-19 impact analysis by offering (printer, material, software, service), process (binder jetting, direct energy deposition, material extrusion, material jetting, powder bed fusion), application, vertical, technology, and geography – Global forecast to 2026", accessible at: https://www.marketsandmarkets.com/

Mckinnon, A., Yongli. G., (2006), "The potential for reducing empty running by trucks: A retrospective analysis", International-Journal-of-Physical-Distribution & Logistics-Management, vol. 36. pp. 391–410. 10.1108/09600030610676268

Merrild, H. et al. (2012), "Assessing recycling versus incineration of key materials in municipal waste: The importance of efficient energy recovery and transport distances", Waste-Management, vol. 32, pp. 1009–1018.

Michelle, J. (2018), "Interview with UPS: How will 3D printing impact the supply chain?", 3D-natives, https://www.3dnatives.com

Mohtashami, Z., Aghsami, A., Jolai, F. (2020), "A green closed loop supply chain design using queuing system for reducing environmental impact and energy consumption", Journal-of-Cleaner-Production, vol. 242, p. 118452.

Mourdoukoutas, P., (2015), "How 3D-printing changes the economics of outsourcing and globalization", Forbes, https://www.forbes.com

Nikolaoua, I., Evangelinos, K., Allan, S. (2013), "A reverse logistics social responsibility evaluation framework based on the triple bottom line approach", Journal-of-Cleaner-Production, vol. 56, pp. 173–184.

Özkan, K., Ergin S., Işık Ş, Işıklı I., (2015), "A new classification scheme of plastic wastes based upon recycling labels". Waste Manag. 2015 January, vol. 35, pp. 29–35.

Peeter, J. et al. (2017), "Economic and environmental evaluation of design for active disassembly", Journal-of-Cleaner-Production, vol. 140, part 3. pp. 1182–1193.

Peng, P., et al. (2020), "Life cycle assessment of selective-laser-melting-produced hydraulic valve body with integrated design and manufacturing optimization: A cradle-to-gate study", Additive-Manufacturing, vol. 36, p. 101530.

PlasticsEurope Market Research Group, (2015), "World plastics production 1950–2015", Plastics-Europe, https://committee.iso.org/

Ray, I., (2011), "Impact of population growth on environmental degradation: Case of India", Journal-of-Economics-and-Sustainable-Development, vol. 2, Issue 8, 2011 pp. 72–78.

Ribeiro, I. et al. (2020), "Framework for life cycle sustainability assessment of additive manufacturing", vol. 12, Issue 3, p. 929.

Santander, P., et al. (2020), "Closed loop supply chain network for local and distributed plastic recycling for 3D printing: a MILP-based optimization approach", Resources, Conservation and Recycling, Resources, Conservation and Recycling, vol. 154, 2020, p. 104531.

The 3D Printing Solutions Company, (2014), "Jigs And fixtures: More profitable production", Stratasys, accessible at: WP_FDM_JigsAndFixtures_0316a_Web.pdf

Wiesmeth, H., (2020), "Implementing the circular economy for sustainable development", Elsevier.

Wong, K., Hernandez, A., (2012), "A review of additive manufacturing", ISRN Mechanical Engineering, vol. 2012, Issue 4.

World Bank Population Density, (2018), can be accessed at: https://data. worldbank.org/indicaton

Wu, H. et al. (2022), "Additive manufacturing of recycled plastics: Strategies towards a more sustainable future", Journal of Cleaner Production, vol. 335, p. 130236.

Zahnd, P., (2018), "3D printing in the automotive industry", 3D Printing Industry can be accessed at: https://3dprintingindustry.com/news/3d-printing-automotive-industry-3-132584/

Chapter 3

Applications of 3D Printing in Product Safety

C.H. Li, T.T. Lee, S.L. Mak, W.F. Tang, and C.C. Lee

CONTENTS

3.1 DEVELOPING 3D PRINTING ALERT SYSTEM IN CONCEPTUAL DESIGN

3.1.1 3D printing model in conceptual design of new product development cycle

Three-dimensional (3D) printing is also called rapid prototyping because the prototype could be built within a day or just a few hours. Conventionally, when we want to build a prototype, we must provide the drawings with detailed dimensions to the manufacturer first. After that, the manufacturer

DOI: 10.1201/9781003306238-3

must find suitable materials for production. Finally, the workers could start to produce the prototypes. Since this is not for mass production, the cost is high and not all the manufacturers are willing to produce a few prototypes using a lot of resources and laborers. Besides, the whole process could be prolonged by the logistics or unexpected delay. The traditional way is also inflexible. The design is likely to be changed for multiple times in conceptual design. It is not feasible to wait for the manufacturer every time when the design is changed. In addition, not all the product design companies have their own workers that could produce the prototypes. Therefore, 3D printing plays an important role in product design. The design team could simply import the 3D model to the 3D printer and the prototypes could be built automatically within a day. This could also provide the flexibility of the design because it would not waste a lot of time to build the updated model when the design is changed. This provides a low-cost and effective solution to the product design team to build prototypes for product evaluation.

3.1.1.1 3D printing model development

a. Look-like model review (aesthetic review for marketing review)

A 3D printer can be used to produce a look-like model, which is a prototype that has the similar appearance to the actual product. It is unnecessary to include all the components of the product in the model. Simply including the body case or other accessible area of the product is good enough for evaluation. This could save the materials and time to build the model. With a look-like model, the marketing team could do a survey to collect user's feedback on the product. Based on the feedback, the design can be adjusted accordingly to satisfy most of the customer's expectations when the product is available on the market.

b. Work-like model review (function review for production)

Work-like model is a prototype that has the function of the actual product. It could be built by a 3D printer to evaluate the behaviours of the mechanical components. The components with different dimensions or shapes could be prepared first. By replacing the components in the work-like model, the design team can find out the best combination of the components that could satisfy the functional requirements. A 3D printer provides a low-cost, flexible and rapid way to design the products. The cost of 3D printed materials are relatively low when compared with metal. During the trial-and-error phase, the design is changed frequently. A 3D printer could promptly help to print a new component but not involve other parties; and could speed up the product design process.

c. Safety model review for hazard review

Safety is important in product design. Products must comply with the safety standards in different countries so that they can be exported. A comprehensive safety model is required during product development because all aspects of the products must be analysed. This model could be produced by a 3D printer for analysing the potential hazards of the product. This could help to find the dangerous parts of the products such as sharp edges or corners. These could lead to injuries when consumers are using the products. For example, the accessible moving part of the interactive part of the product may trap the finger and cause injury. After finding all possible hazards, the design team could adjust the design and eliminate the hazards

In product design, aesthetics is one of the major concerns to attract consumers in marketing review. The attractions representing the emotional response are related to novelty and pleasant traits. The feeling of product aesthetic by consumers are affected by moderators such as quality of manufacture, brand reputation or viewing angle (Wu et al., 2017). Product aesthetic is subjective to personal judgement and individual preference. It is defined as how a product looks, feels, sounds, tastes or smells. Unlike other dimensions, aesthetics is intangible and cannot be represented by a numeric value. It is ineffective to use non-tangible analytical prototypes to do the evaluation in a mathematical or visual manner because we are not able to touch it. Therefore, in the early design stage, it is difficult to evaluate the product aesthetic until the product is available on the market. However, it is risky to wait for the response from the consumer after the product is sold. This will also affect the brand reputation mentioned previously if the product review is negative. This could be solved by evaluating the product aesthetic in the early design stage.

The core element of the products is the function. When designing the mechanical components, it is difficult to interpret whether they could work as intended based on the virtual 3D model. Dimension is one of the concerns in the function. The value of length, width and depth could have different combinations. Different combinations of dimensions of the components could lead to different effects on the function. Although this could be done by computer-aided engineering (CAE) to simulate the stress distribution which helps to find the optimum dimensions, the actual movement of the components is still unknown. CAE could only help to narrow down the dimension range. The optimum dimensions must be obtained based on different experiments. Force is also a design concern in the function of the product. Many mechanisms are using friction to achieve the function. Designers must design an effective shape or pattern to ensure the friction is enough to support the action.

Design for assembly (DfA) focuses on ease of assembly of the product and minimising the complexity of the assembly process. However, 3D printing models are always limited by size. A typical 3D printer can only produce small parts. In order to produce a model in which the size is over the limit of a 3D printer, the model must be separated into smaller parts. Later, to assemble the parts produced by a 3D printer, screws are commonly used. However, it is not feasible for a 3D printer to print a screw thread for a small hole because of the tolerance. The support will also affect the quality of the screw thread. Therefore, generally there are two methods to solve the problem. The first method is to print a hole that is slightly smaller than the diameter of the screw. For example, for a M4 screw, a hole with a diameter of 3.3 mm will be used. However, the difference between the hole size and screw diameter depends on the tolerance of the 3D printer. This value could be found out by doing several experiments. Since the hole size is smaller than the screw, the screw could be used to tap a hole so that screw thread is created since the materials used by 3D printers are usually soft. However, this is not a reliable method. Therefore, in second method, a square hole that is perpendicular to the screw hole will be added to the model. The square hole allows us to insert the screw nut so that we can use the screw thread of the screw nut to assemble the two components. This simplifies the assembly process, and the assembly is more reliable. This method is also easy for maintenance and replacement when compared with the previous method. However, more effort is required to add the square hole to the model.

Design for manufacturing (DfM) focuses on minimising the cost and time of the production process while maintaining the quality of the product. Using a 3D printer could save a lot of time because the product could be obtained directly without having to wait for the courier if the manufacturing company has their own 3D printer. However, the quality of the product would be a concern. Usually, 3D printed components are not durable which will be worn out or broken easily. In addition, the strength of the components is not strong enough to resist a large amount of external force. Therefore, not all situations could apply to 3D printed components. It is not recommended for permanent use or to resist force.

The costs are low when compared with producing metal components. The cost of 3D printed materials are typically low. Each pack of the materials is enough to support the consumer products. On the other hand, traditional production methods require workers to mill the components. The labour costs for making a single component would be very high, and usually it contributes to a large portion of total cost. However, 3D printers could produce the component automatically without extra workers, this could significantly save the manufacturing cost.

3.1.1.2 Rapid changes of 3D printing model

There are three major versions in the conceptual design phase. The first version is the look-like model for marketing that is reviewed by the marketing team or licensor. The second version is the work-like model for mass production that is reviewed by the manufacturer. The third version is the safety model for hazard review by a quality assurance team or commercial laboratory. For each version usually several sets will be prepared. A 3D printer could provide the flexibility to adapt to the rapid changes of the model.

3.1.1.3 Waste of 3D printing model

Stringing is a frequent problem in 3D printing. As a result, many filaments would appear on the surface of the components. The simplest way is to reduce the nozzle temperature in order to increase the material viscosity so that stringing is minimised. This is because when the temperature is high, the materials would be more liquified and extra materials would drop onto the components at the end of the path. However, the temperature should not be set too low or otherwise the material cannot be extruded. In addition, increasing the extrusion speed could also solve the problem. However, an optimum speed should be found first. If the speed is too high, the materials printed onto the surface may not be enough. The settings of the parameter are based on the material to be printed since physical properties vary for different materials.

The improper settings are critical to the look-like model. The rough surface of the prototype will affect the judgement on aesthetic. Besides, stringing could not be eliminated in some cases even with optimum setting. For example, when thin or small objects such as words or logos are printed, stringing is likely to occur. Therefore, those are not suitable for aesthetic review. In contrast, large objects with less complex shapes such as body cases could be used for aesthetic review.

3.1.2 Design alert system in conceptual design

3.1.2.1 Product recall in the market

Product recalls can be devastating to different stakeholders in a supply chain. There is an estimate of about 29,000 casualties in the United States every year from consumer product-related injuries. Together there are also approximately 33 million non-fatal consumer product-related injuries (U.S. Consumer Product Safety Commissions, 1985). An annual economic loss for more than 1 trillion dollars is caused by deaths, injuries and property damages from consumer product accidents (U.S. Consumer Product Safety Commission, n.d.). An independent commission should be crucial to monitor and provide standard guidance for designing consumer products, in order to improve product safety and reduce the risks of related injuries.

The U.S. Consumer Product Safety Commission (CPSC) is established under the Consumer Product Safety Act (CPSA) in 1972 by the Congress. The CPSC is highly recognised as the global leader in monitoring the safety and potential hazards of consumer products. (U.S. Consumer Product Safety Commission, n.d.)

The CPSC has lowered consumer product-related moralities and injury rates over the past 50 years through its services and contributions. It develops voluntary standards with standards organisations and different manufacturers and businesses. The consumer products are banned in case if there is no workable standards to adequately ensure the safety of the users. Since 1994, it has negotiated on more than 145 voluntary safety standards and has published more than 40 mandatory safety standards. It is also responsible for making announcements in public, when it may offer up to arranging for the replacement, fixing or refund for the recalled products and identifying the product hazards, when necessary. It works to advise and educate consumers, manufacturers and other stakeholders in the supply chain, through online seminars, traditional workshops and social media. (U.S. Consumer Product Safety Commission, n.d.). Every year, it reviews about 8,000 accidental product-related casualties across all 50 states and the District of Columbia. It achieves an annual saving of about $13 billion in property damages, and health care and other social costs (National Electronic Injury Surveillance System, n.d.).

3.1.2.2 CPSC product safety recall

In case if a substantial hazard to the public is identified in an existing consumer product, the CPSC, empowered by the Consumer Product Safety Improvement Act (CPSIA), possesses the authority to initiate the corresponding product recall. It can also raise the civil penalties in case of product safety violations, and set up mandatory performance standards to certain tests and certification requirements, warning and instruction requirements, for consumer products.

Upon identifying an intentional violation against Section 19 under the CPSIA, the CPSC may seek U.S. courts to impose civil penalties to the corresponding company. A violation can be any of the cases like: failing to report potential product hazards; not meeting the Commission Standards as required by Law; sales of recalled products and unauthorised use of certification mark.

Data from recall reports makes the information about the recalled products open for download, including the descriptions and photos about the potential hazards and the corresponding manufacturers. The data is therefore very useful and valuable to product designers, who can learn about the identified potential hazards in similar products. Review over design problems and over previous recalls from such reports can therefore be possible. Such data provides factual supporting references to product designers to make evaluations and decisions, and strengthens them on top of their own knowledge and experience on product safety. The product designers can proactively use the CPSC product safety recall data, to modify the product design before printing a 3D conceptual model and before mass production, to prevent potential safety hazards.

3.1.2.3 CPSC national electronic injury surveillance system

This data is extracted from the CPSC activities, being a statistically validated injury surveillance and follow-back system and gathers from U.S. hospital emergency departments about injury data related to consumer products in the United States. It works on a probability sample of various scales and locations, to provide a countrywide estimation and types of injuries.

Every year, National Electronic Injury Surveillance System (NEISS) collects information on almost 400,000 product-related cases from the sample base (being collected from about 100 hospitals). The hospitals transmit timely event information electronically, when the data can sometime be available in less than 24 hours after the event (The United States Consumer Product Safety Commission, 2017).

The CPSC publishes an annual NEISS Data Highlights report, which summaries injury data by major product categories, and by age and gender. This report is a good entry point to review the common problems on product design and safety.

Product designers can pinpoint their inquiries under the NEISS database through specifying one or more variables. Potential hazards of different consumer products can be quantified and estimated. NEISS has become a very important product safety research tool for product designers throughout the world.

Other countries also established similar systems to collect safety data and publish alert systems on consumer products. In the European Union, for examples, RAPEX is the Rapid Alert System for dangerous products (but not including motor vehicles and the recreational craft); RASFF is the Rapid Alert System for Food and Feed (but not including food packaging). Different systems work in specific areas to monitor the product safety.

3.1.2.4 Design database

To increase the efficiency of the design process cycle and eliminate the probability of product recall, creating a 3D printing database would assist the product designer proactively. With the establishment of a 3D drawings database, it centralises the drawings and evaluates the design based on product safety. Based on the drawings on the software, a safety checklist and safety alert could be provided through analysing the common risks of different categories of product. For example, if the designer of the consumer product has a potential sharp point and sharp edge hazard, the system can pop out a safety alert to remind the designer to bear in mind certain potential hazards. The design database can be based on the CPSC data and update regularly.

3.2 STANDARDISING 3D PRINTED PRODUCTION TOOLS

When production is running, mass quantities of materials, components and production tools should align with the tightened product-launching schedule. Companies usually assign their manufacturing plants in more than one country for production across different regions at the same instant. Product quality would be a major concern as these may be inconsistent as the deviations of the production tools used in different manufacturing plants.

Once the conceptual design is finalised, companies could authorise one of manufacturing plants to fabricate 3D printed production tools for conducting a trial production. They can then check any problems in production line, product safety and other uncertainties; and further amend the corresponding problems and provide the enhanced digital 3D modelling file for standardising the design of production tools from one production plants to the others in order to minimise any deviations or errors of production tools. Hence, the product quality and consistency will be secured.

The major applications of production tools are in sample assembly process and sample inspection (MacLean-Blevins, 2017).

a. Production assembly process

During sample assembly process, there are different stages which test fixtures are used. A plastic fixture is used to burn firmware programme to micro processing unit (MPU) on printed circuit board assembled (PCBA). This fixture connects golden fingers on PCBA and connect to a PC through flat cables (Alitools, n.d.). Besides electronic assembly plants, fixture can be used as a support on silk screen printing of round plastic cases. Plastic or plaster fixtures are used to hold or fix the position of the product for in the assembly process (AliExpress, n.d.). Other sample assembly processes, including ultrasonic welding, wire routing and wire soldering, may include the use of fixtures or jigs to support the completion of manufacturing process in a production plant.

b. Product Inspection

Test fixtures and test jigs are also used in sample inspection process. One example is the test of PCBA functions after its manufacturing process in PCBA manufacturing plants. In this process, a test fixture is built with the installation of input and output devices such as LCD display, push buttons and buzzers. When a PCBA is manufactured, it is placed into the fixture. This fixture supplies power from a DC power source, and functions of the PCBA is checked with push buttons and displays on the test fixtures. The test fixtures are specially built for each PCBA and is traditionally made by acrylic plastic (Royal Circuits, 2020).

Other test fixtures are built for quality assurance process, such as test fixtures for durability test of hinges, and fixtures for electrical safety tests such as hi-pot checking and burn-in systems for switching mode power supply or light bulb manufacturing plants.

3.2.1 Common types of test jig and fixture in the manufacturing industry

In product assembly process, test fixtures are used to assist in fixing the position of the product supporting the installation. For example, test fixture is used to assist in fixing positioning of ultrasound welding process, and silkscreen printing process of plastic cases. Traditionally, there are three common ways to prepare test fixtures. With the price from low to high, they are plaster molding, acrylic fixture and metal test fixtures.

a. Plaster-molded test fixture

Plaster-molded test fixture is a low-cost solution to build a test fixture from plaster. The plaster mold is made with the real product parts. Plaster mold formed by pouring liquid plaster into a container fitted with real product parts (usually the case) as a mold. After plaster is dried, the fixture could be used to assist in production process. Like fixing positions for further manufacturing steps. Although the cost of this plaster mold is the lowest among the 3 common types, it is not durable, especially in heavily loaded manufacturing plant. Moreover, it is not used for production steps which precision is needed.

Despite its low cost, plaster-molded test fixture is brittle and with low precision. Plaster test fixture shall be replaced frequently in a production line. In addition, with low precision of this type of test fixture engaged, the frequent altering of test fixture introduced quality issue in terms of homogeneity.

Plaster-molded test fixture is waring with respect to the time it is used. This changes the dimensions of the fixture, thus affects the accuracy in some manufacturing processes. Furthermore, plaster-molded test fixture's precision is highly affected by the humidity and temperature of the production plants. For production plants located in different countries, the durability, replacement and maintenance rate of these fixtures vary. This introduced great uncertainties in the manufacturing process in the world of globalisation.

With 3D printed test fixture, the selection of material allows manufacturer to make a balance between durability, precision, and cost. Three-dimensional printers can make models from polyvinyl chloride (PVC), acrylonitrile butadiene styrene (ABS) or even metal. Even for the lowest cost PVC fixture, the physical durability is better than fixture made from plaster.

In addition, plaster molds have physical differences from one manufacturing plant to another. In contrast, 3D printed test fixtures are based on electronically transferrable 3D drawing files. The file is identical independent to plants using it. The precision is controlled by the 3D printers. In theory, with the same precision given across similar models of 3D printers, the test fixture is literally identical in most manufacturing plants in practice.

b. Acrylic test fixture

Acrylic fixture is a medium-cost solution to build a test fixture based on acrylic plates and metal parts. This type of test fixture is commonly used in electronic assembly lines fixing electronic parts such as LCD screens, and connector components. The preparation of acrylic test fixture usually involves a third party to support the building of such fixture. With an experienced fixture production party, the drawing of the fixture is designed. Then, it is built manually by cutting acrylic sheets from its raw material. Then pieces of cut acrylic components are glued and assembled. With hand drilled screw holes, metal screws, hinges and levers are assembled to form an acrylic test fixture. This test fixture provides better durability than plaster fixture, and more precision in the manufacturing process.

While acrylic test fixtures are giving higher durability in the manufacturing plants, the creation of this fixture type needs to employ experienced staff or third party in design. Acrylic test fixtures are not built directly from 3D model files. The fixture preparation process includes cutting pieces of acrylic boards from huge sheets of acrylic plates. Boards of acrylic plates have fair accuracy on rectangular pieces, while for special shaped boards, the accuracy subjects to several criteria like usage of automatic tools, manpower skill and knowledge level, and other environmental factors. Since the boards are prepared by cutting out the unwanted material, a lot of acrylic material is wasted in the fixture preparation process, and this will be reflected indirectly in terms of costs of fixture.

After cutting the acrylic boards, these boards are glued together by craftsmen. Again, the precision of this fixture varies depending on the skill level of craftsmen, as well as the time available to build the fixture. Also, the labour costs of craftsmen varies with his/her skill level.

Another major drawback of acrylic test fixture is the production time. Due to the need of assemble of the test fixture and preparation of material, complex test fixtures may take seven to ten working days of preparation. With a huge scale of production line, the time cost of test fixture increases while the precision of the test fixture remains at a similar level in the manufacturing process.

With 3D production tools are built from a 3D printer, 3D printing process is a material addition process. Therefore, it minimised the waste of material from plastic boards preparation. With previous discussion in the building process from electronic 3D model files, there is little deviation of the models built from the 3D printing process.

Three-dimensional-printed test fixture can give a higher durability as compared with acrylic fixtures. For plastics with better durability, acrylonitrile butadiene styrene (ABS) can be used to give a satisfying level of durability for the manufacturing process.

The preparation time of 3D fixtures are much shorter when compared with acrylic test fixtures. A palm sized test fixture with assemble process needed can be built in two working days (MacLean-Blevins, 2017), which is five times more efficient than that of the acrylic test fixtures. Moreover, the skill set needed to build 3D printing fixture with its assembly process is much lower than that of the production of acrylic fixtures. For smaller manufacturing plants, there is no need for them to pay for external parties to build such test fixture. These plants can bring the fixture construction process in house by introducing a 3D printer with limited price. The overall fixture costing could be saved.

c. Metal test fixture

Metal test fixture is a fixture used for plastic molding, and heavily used scenarios (adapted from Tension Technology, n.d.). Metal fixtures are prepared by metal workshops through standard metalworks such as assembling, cutting and bending. Like acrylic test fixture, the preparation of metal fixture shall be handled by specialised craftsmen at metal workshop, usually outsourced. The design of metal fixture is provided by product engineer with review and modification by experienced craftsmen in the workshop. This kind of test fixture is the most durable and can be used in assembly processes which high pressure or forces is engaged to hold the product in place.

Traditional metal test fixture, despite the durability of the fixture, is the most expensive type of test fixture as high material cost and fabrication technique are required. The preparation of test fixture consists of turning, welding and different metal works. Its accuracy is largely relied on the skillfulness of the party who is building the fixture. The increase in the precision requirement of the fixture results in a dramatically rise in the cost and a prolonged time to prepare for the fixture.

The preparation time of this fixture is also the longest among the three types of production and test fixtures. This type of test fixtures can hardly be mass produced, and there is a high potential for small- and medium-sized manufacturing plants ordering tailor-made test fixtures from a third-party fixture producer, which is another significant budget added to these enterprises.

Since most test fixtures are built by third-party metal work production houses, there might be communication error between manufacturing plants and metal work production houses. For metal test fixture, even with small deviation of understanding during the metal test fixture preparation process, the fixture built is often irreversible. This implies a small communication error would result in reworking on the whole test fixture, resulting in huge time and costing lost.

Latest 3D printing technology has already included metal, such as steel and copper, as a printing material. This gives an alternative option for the manufacturing process which high level of durability is needed in the test fixture during the manufacturing process. With precisive 3D printers available in the industry, the accuracy of a typical 3D printed test fixture can be minimised, with an additional uniformity proofed across different manufacturing plants.

Three-dimensional-printed fixture does not require craftsmen with high level of metal work skill sets. The quality of the fixture is controlled by the 3D printer, which is already a mature technology in the domain of industrial manufacturing. The time of preparation can be shortened to two days from weeks of preparation needed for external test fixture manufacturing parties.

Three-dimensional-printed fixture are prepared based on the design of a 3D model generated by design engineer. The electronic files of the models will not be different across different parties. The printing by the 3D printers will work according to the 3D model file. Therefore, the risk of communication error between fixture production and manufacturing session is minimised.

3.2.2 How 3D printed test fixture provides the best replacement for traditional test fixtures?

With the introduction of 3D printed test fixture, test fixture can be built using 3D printers of different printing materials like metal powder and plastic filament. The durability and cost can be optimised with the selection of suitable material to build the fixture and the cost of the 3D printing machine. For instance, when compared a plaster fixture with a 3D printed ABS fixture, the 3D printed fixture would outperform the plaster one no matter in durability and cost.

About the quality and uniformity of the test fixture, since fixtures are 3D printed, the quality is uniform across different production plants. The same fixture can be built without the consideration of skill level of craftsmen in the workshop. This can be applied on metal sheet banding process.

The design of the test fixture can also be simplified by the reuse of the 3D drawings of the product to give the best dimensions in the design of the test and production fixtures. With affordable 3D printers, fixtures are no longer required to be built by external parties with a long period of logistic time and procurement processes. The turnaround time of the production of test fixture can be greatly shortened.

The precision of the 3D printed fixture can be highly controlled. With 3D printing technology, there is no longer a need to manually cut acrylic and metal sheets and weld or glue them up during the fixture production process. The positions of screw holes and hinges are controlled by the 3D printer which precision is guaranteed and well specified by the 3D printer. This gives 3D printed fixture outranged workshop-built fixture in terms of stability and precision.

3.3 CENTRALISING 3D TESTING IN THE TESTING, INSPECTION AND CERTIFICATION INDUSTRY

Testing, inspection and certification (TIC) refers to testing, inspection and certification industry. For testing, usually a product is tested by a commercial laboratory according to the international or national standards. A testing report will be issued to indicate whether the product satisfies the requirements. For inspection, a product is checked based on professional judgement. For certification, a management system or product will be audited by third-party certification bodies. The TIC industry plays an important role in conformity assessment so that customers or the environment are protected.

According to the final report from the TIC Council in 2020, the current global market size is estimated to be US$200 billion. It is expected the size will be over US$260 billion by 2025. Outsourcing of TIC services has a growing trend in the industry. There are around 1 million employees from more than 160 countries in the TIC industry to provide conformity assessment activities. The TIC industry is a large and growing sector (Europe Economics, 2020).

There are approximately 10,099 (IBIS World, 2021b) and 2,855 (IBIS World, 2021a) of laboratory testing services in the US and Canada, respectively. According to the statistics in Hong Kong Council for Testing and Certification, there were 830 testing laboratories in Hong Kong in 2018 with 18,690 of people engaged (Yim, 2020). Based on the laboratory to number of people engaged ratio, for one million people engaged in the TIC industry around the world, it is estimated to have over 40,000 TIC laboratories in the world.

When a product requires conformity assessment service, the company will seek for consultation from a commercial laboratory first. Different countries have different requirements and standards. The countries to be exported need to be determined first. Some countries like the US or EU countries have strict rules and regulations on environmental protection. Apart from safety and functional requirements, environmental standards should also be complied. The product should conform to all the mandatory standards or otherwise cannot be exported to those countries.

After consultation, the laboratory will confirm the standards to be applied. The testing method to be conducted according to the national or international standards is called standard methods. If the testing standard could not fully satisfy the requirements, the laboratory could provide an in-house method based on the requirements from the client. In-house method is a testing method that applies the procedure of standard method but with modified criteria. Some clients might use an in-house method as they would like to have stricter requirements to ensure the quality of the product. There are also non-standard methods that are developed by the laboratory for special products which are agreed or requested by the client. Method validation and verification should be conducted according to the procedure and requirement specified in international standard ISO/IEC 17025.

After confirming the testing method to be used, the commercial laboratory will select an appropriate testing machine and tool to test the product according to the requirement specified in testing standards. Some typical machines such as the universal testing machine (UTM) are commonly used in the industry to test the tensile strength and compressive strength of the product. To test the strength of the product, one end must be fixed as the reference point so that the stress could be measured. Different shapes of product would require different kinds of fixture for example V-block will be used for holding the cylinder-shaped product.

Testing will be conducted after the testing setup is completed. In general, testing could be conducted automatically or manually. For some tests that require long duration or many repeated cycles, it is not possible to require the technician to do the test repeatedly for long hours. For this kind of test, an automated testing machine is usually available. Once the technician finishes the setup, the machine will do the test automatically until the end of the testing process. However, some monitoring is required because the machines sometimes give an alert and the test will be paused. For manual testing, usually the tests are short and simple which does not require a machine to test.

3.3.1 Existing practice of testing machine, tool and fixture in the TIC industry

In the TIC industry, although manufacturing stakeholders are following the testing standards for testing the consumer products in different stages of the new product development process, most of the testing machine, tool and fixture are designed and made by individual manufacturing stakeholders. A testing machine with the same function could vary in structure, features, dimensions, tools and fixtures for different stakeholders. It would be complicated to verify all the testing machines made by individual manufacturers.

a. Purpose of making testing machine, tool and fixture

Not all tests could be handled manually. Therefore, testing machines plays an important role in the TIC industry. Some tests require steady and stable force or movement applied to the product. This could not be done manually because technicians could not control the amount of force applied to the product precisely. It is also impossible for technicians to apply the exact same force to the product in each cycle. To ensure the quality of the testing process, a testing machine is required to provide the reliable result. The testing machine will also have some accessories such as a testing tool that help technicians to do the testing setup. A fixture could be a component of the machine that is replaceable. This could ensure the accuracy of the measurement or otherwise the value measured would be underestimated since not all the forces are applied on the product.

For a commercial testing laboratory, several aspects needed to be concerned before purchasing the testing machine. The machine should be able to conduct the test that fulfills the requirements of the testing standards from different countries. Different standards will have different requirements on the parameter setting. The machine should cover a wide range of testing conditions. In addition, safety is also a major concern when purchasing a testing machine. Safeguards or devices should be provided and comply with the local safety law to ensure the safety when operating the machine. Otherwise, the laboratory must make their own safety devices before operation. Thirdly, the stability of the machine is also a key factor. A machine should be automated without frequent monitoring. Machine downtime should be minimised as much as possible. The machine should also list out all the preventive measures that would affect the operation of the machine.

For manufacturers, they will have difficulty due to lack of resources and technology. In order to lower the manufacturing cost, they tend to use testing machines with the old technology without upgrading. These machines may have a risk of not complying with the existing testing standards.

b. Common problems on testing machine, tool and fixture

i. Dimension deviation

In general, the testing standard does not specify the details of the testing machine. The individual stakeholders will make the machine based on their own experience. Therefore, different machines produced by different stakeholders will have deviation in the dimensions. The testing quality would be inconsistent.

ii. Long time

Individual stakeholders have less support from other partners. It would take a lot of time to design and manufacture the machine. If the machine fails to meet the requirements, the design needs to be reviewed and modified which will further take a lot of time. The time for logistics and customs is also a concern because there are many uncertainties such as political issues blocking the machine from entering the country.

iii. Insufficient development parameter

The resources of individual stakeholders are insufficient. All the features must develop on their own. It would be difficult to make a reliable machine that is up to standard as well.

3.3.2 Centralising 3D printing file in the TIC industry

Universal testing machines are commonly used in the TIC industry. However, the shape of the products is not standard or sometimes irregular. If the laboratory does not have a suitable fixture to hold the product, they must design a new fixture and find a manufacturer to produce the parts. Without enough experience, the design might need to be modified several times and it will take a long time to make a new fixture. If all the fixtures are centralised in the cloud server, the laboratory could download and produce the fixture immediately. Inside the shared storage, the folder should be sorted by different categories so that the user could find what they needed easily. Also, there should be sub-categories for the product with different dimension ranges. For each design, the source drawing file should be provided so that the user could modify the components easily since not all the design could cover all the needs from the users.

The principle could also be applied to multinational corporations. The company may have factories in different countries. The fixture from the outside may not be suitable for the company's product. Therefore, the headquarters of the company could design the fixture and distribute it to the factory in different countries.

a. Minimise dimensional deviation of testing machine and tool

When the users around the world use the same design of testing machine and tools, the dimensional deviation could be minimised. This could help to ensure the testing quality. The results are more consistent and trustworthy. Currently, testing reports issued by developing countries might not be reliable. With standardisation, as long as the testing method is validated, the test result conducted by different countries should be consistent and reliable.

b. Distribute the standardised 3D printing file to global TIC laboratories and manufacturers

Cloud services such as AWS or Azure are useful tools for centralising the testing machine and tools. Every stakeholder could download and upload their file to the shared storage easily. This could help to standardise the machines all round the world. If the TIC council does not want to upload the file to third-party organisations, they could deploy an on-premises server. The file should be stored in the Network-attached storage (NAS) with the protection of a redundant array of inexpensive disks (RAID) to prevent losses. However, extra cost is needed. A web portal should be created so that the member could access the content easily. The global TIC laboratories and manufacturers should be members of the standardisation council so that they could have an account to login and access the content.

File extensions with.stl. and .STEP should be available for 3D printing and modifying the model respectively.

c. Advantages of centralisation of 3D printing file

i. Align testing instrument and calibration

With a standardised testing machine, the calibration procedure could also be standardised. In the past, when calibration was required, the laboratory had to find the supplier of the machine. However, if the supplier no longer exists, it will be difficult to find an expert to calibrate the machine.

ii. Enhance testing reliability

The testing result would be more reliable. The testing quality should be consistent which is independent of the machine supplier or the country that carries out the test.

iii. Shorten development lead time

There is no need to spend time on designing new components. The components could be produced on-site immediately without waiting for the delivery. The whole development schedule for project management could be controlled easily since no third-party is involved.

iv. Reduce any argument of interpretation of testing method

When the testing results are produced by the standard machine that using the recognised testing method, the chance of argument could be reduced.

3.4 ONLINE 3D PRINTING HUB

The following sections discuss how online 3D printing hub can be incorporated into different scenarios.

Traditionally, for example, when a consumer purchases an electronic appliance from the store and some defects on the product are later discovered, this usually results in the need to recall the product or to send the product back to the manufacturer to replace the faulty parts. This is time consuming and causes inconvenience as the consumer needs to bring the product back to the manufacturer collection points or retail stores for product recall. With the 3D printing technology, the consumer can possibly replace the faulty part on their own without having to send the product back to the manufacturer. One of the ways this can be done is as soon as the manufacturer has identified the problem, they can redesign or modify the component and release a video to educate consumers regarding the faulty part. With the assistance of 3D printing, consumers can "print" the faulty component and replace it with a new functional part on their own. The following parts discuss more about the advantages, disadvantages of 3D printing and various scenarios in which 3D printing can be applied to replacing faulty parts in consumer products.

Three-dimensional printing is not something new, as it has been available for a few decades and has become much more popular in recent years as the prices of the 3D printers have significantly reduced. There are several 3D printing methods, namely Selective Laser Sintering (SLS), Stereolithography (SLA) and Fused Deposition Modelling (FDM).

a. Selective Laser Sintering

In this method, a tiny SLS particle is distributed on the surface of the platform and scanned by laser beam to form the first layer. Afterwards, the printing platform is adjusted to lower level and the foresaid SLS process is repeated to form the other layers until the 3D product is finally formed. This process is more flexible as it does not require support material and suitable for printing different materials. By developing and using inkjet printers and binding agents in the binding process (rather than lasers), this method is capable of full-color printing for the material (Brooks et al., 2014).

b. Stereolithography

SLA is a process that uses light exposure to solidify a liquid resin. The advantages of this process like supporting frame (Brooks et al., 2014). A movable platform is kept lowing for printing and lasing the resin until 3D printed product is formed.

c. Fused Deposition Modelling

This is an extrusion operation, where a filament wire of printing material is put into the preheated nozzle. The material is heated up by the nozzle and a series of motors control the liquefier head and extrusion nozzle for making model. Recently, FDM has been applied for making consumer products like footwear in the consumer product industry (Brooks et al., 2014).

FDM is applied in 3D printing technology due to its simplicity and flexibility. Desired shapes could be built by extrusion operation with short operation time and low cost by comparing with other 3D printing methods. Due to the competitive price, this is popular method in the market nowadays (Hwang et al., 2014). FDM also known as personal 3D printers (Ćwikła et al., 2017).

3.5 ADVANTAGES OF 3D PRINTING

By comparing conventional production, 3D printing offers low cost for customising the product and minimising time that is needed (Khajavi et al., 2014). This includes improving the manufacturer's ability to address the safety problem immediately. Using this technology can reduce the lead time needed for corrective action to be carried out by the manufacturers. Furthermore, using 3D printing for replacing faulty parts can diminish delivery cost of new components from the factory to retails stores and ultimately to consumers. Another advantage of using 3D printing technology is that it can enhance the interest in the product by consumers, as they can easily "print" a new component and replace it on their own.

3.6 DISADVANTAGES OF 3D PRINTING

In some cases of 3D printing, such as printing metal, the fabricated items can be prone to shrinkage, brittleness and uncontrolled porosity. There is also the challenge of setting parameters such as deposition speed and retraction speed before printing, which directly impact the 3D printing surface and structural integrity (Kantaros et al., 2022).

Certain problems found in FDM 3D printers may include motor stall due to excessive impact or resistance, nozzle blockage because the material is not melted and abnormal wire extrusion caused by motor abnormality (Liao et al., 2019).

In the following part, we will discuss four different scenarios involving how the consumers can obtain the 3D printed component.

Scenario 1: Consumer downloads the 3D printing file and print it by themselves at home.

The main advantage of this scenario is that it is all Do-It-Yourself (DIY), the consumer can do everything on their own at home and can obtain the component wherever they want, and can have the faulty component replaced as soon as possible.

The downside to this is the user needs to have their own 3D printer at home to be able to do this. One limitation regarding having the 3D printer at home is the component may be a large in size that requires a large 3D printer to print. Storage can become an issue, as there may not be space to store a large 3D printer at home that is only used sparingly.

Another problem is the 3D printer that the consumers have at home may not be compatible with the 3D printing file provided by the manufacturer, therefore even if they have downloaded the 3D printing file, they may still not be able to print the component at home.

After the consumers have printed the component at home, they may still encounter the problem where they may not have the tools required to assist them with replacing the faulty component.

Furthermore, the consumers may not have the 3D printer that can print the component using the necessary materials. For example, the FDM 3D printers need to use a Metal/Polymer Composite filament for printing, yet this filament may not be suitable for the faulty component. Even if the component is printed, there may be operational hazards if the material used for printing is not the suitable one.

Despite the relative convenience of printing the faulty component at home, there may still be the issue of the consumer not knowing how to replace the faulty part, even if they follow video instructions provided by the manufacturer.

Scenario 2: Consumer collects the 3D printed components from the retail store, then fixes the problem by themselves at home.

The advantage of this scenario is that the consumers have the product printed for them at the retail store and do not have to worry about printer issues or using the wrong materials for printing the component. After the consumer collects the 3D printed part from the retail store, they can replace it at home. The consumers do not have to worry about storage issues regarding the 3D printer at home. Also, consumers do not have to worry about the incompatibility issues with the 3D printing file downloaded and the 3D printer at home.

One main disadvantage of this is the consumer must travel to the retail store to collect the component, which may be time-consuming. Another thing is even if the consumer has collected the component, they may not know how to replace the faulty component on their own at home.

As mentioned earlier in scenario 1, certain specific tools may also be needed for installing the component, whereas the consumers may not have those tools at home, unless the specific tools needed can be printed by the 3D printer together with the component.

Scenario 3: Consumer collects the 3D printed components from the retail store, then fixes the problem by retail shop simultaneously.

The advantage of this scenario is that the consumers have the component printed for them at the retail store and replacing it at the retail store simultaneously. The consumers do not have to worry about not knowing how to replace the component, as there will be assistance from the staff at the retail store. The consumers also do not have to worry about the 3D printer materials and software incompatibility, as all this will be provided by the retail store.

The obvious drawback of this scenario is it is very time consuming. As the consumers need to travel to the retail store which may be inconvenient if they live far away. The consumers not only collect the component from the retail store, but they also then need to spend even more time at the retail store to replace the faulty part. Another downside is the consumers will have to bring the whole faulty product (e.g., electronic appliance) to the store, which may be a hassle, as the product may be heavy.

Scenario 4: Retail store prints the 3D printed components. Consumer collect the printed components from the retail store, then fixes the problem by themselves at home.

Much like Scenario 3, the advantage of this is consumers do not need to worry about the quality of the printed component, as it is provided by the retail store, yet the consumers do not need to physically bring the electronic appliance to the store, saving the hassle. The consumers can then take the component home, and replace the faulty component on their own.

The downside to this is like in Scenarios 1 and 2, the consumer may not know how to replace the faulty part on their own at home, even with video assistance provided from the manufacturer. Also, the consumers may lack the tools needed to replace the faulty component.

3.7 SUMMARY

As 3D printing technology advances, prices of 3D printers have become cheaper and cheaper, therefore it is more feasible and easier for consumers to obtain 3D printers and 3D printed components. As seen from four different scenarios above, there may be limitations and drawbacks with printing components using 3D printing, yet there are also many benefits that can outweigh the disadvantages, hence 3D printing components may be worth exploring in the near future.

REFERENCES

AliExpress. (n.d.). Disposable paper cup silkscreen print machine Guangzhou semi auto.

Alitools. (n.d.). LCD test PCB clip fixture probe spring pin programming burn download 2.54 mm test rack PCB clamp fixture jig probe 15P-20P.

Brooks, G., Kinsley, K., & Owens, T. (2014). 3D printing as a consumer technology business model. *International Journal of Management & Information Systems*, *18*(4), 271–280. 10.19030/ijmis.v18i4.8819

Ćwikła, G., Grabowik, C., Kalinowski, K., Paprocka, I., & Ociepka, P. (2017). The influence of printing parameters on selected mechanical properties of FDM/FFF 3D-printed parts. *IOP Conference Series: Materials Science and Engineering*, 227(1), 12033. 10.1088/1757-899X/227/1/012033

Europe Economics. (2020). *Value of the Testing, Inspection and Certification Sector*. https://www.tic-council.org/application/files/1216/2211/4719/Value_of_the_Testing_Inspection_and_Certification_Sector_-_2020-12-23_Final_report.pdf

Hwang, S., Reyes, E. I., Moon, K.-s., Rumpf, R. C., & Kim, N. S. (2014). Thermo-mechanical Characterization of metal/polymer composite filaments and printing parameter study for fused deposition modeling in the 3D printing process. *Journal of Electronic Materials*, 44(3), 771–777. 10.1007/s11664-014-3425-6

IBIS World. (2021a). *Laboratory Testing Services in Canada –Market Research Report*.

IBIS World. (2021b). *Laboratory Testing Services in the US –Number of Businesses 2002–2027*.

Kantaros, A., Piromalis, D., Tsaramirsis, G., Papageorgas, P., & Tamimi, H. (2022). 3D printing and implementation of digital twins: Current trends and limitations. *Applied System Innovation*, 5(1), 7. 10.3390/asi5010007

Khajavi, S. H., Partanen, J., & Holmström, J. (2014). Additive manufacturing in the spare parts supply chain. *Computers in Industry*, 65(1), 50–63. 10.1016/j.compind.2013.07.008

Liao, J., Shen, Z., Xiong, G., Liu, C., Luo, C., & Lu, J. (2019). Preliminary study on fault diagnosis and intelligent learning of fused deposition modeling (FDM) 3D printer. In 2019 14th IEEE Conference on Industrial Electronics and Applications (ICIEA) (pp. 2098-2102). IEEE. 10.1109/ICIEA.2019.8834376

MacLean-Blevins, M. T. (2017). *Designing Successful Products with Plastics: Fundamentals of Plastic Part Design*. William Andrew.

National Electronic Injury Surveillance System. (n.d.). Introduction.

Royal Circuits. (2020). Royal circuits partners with Ponoko.

Tension Technology. (n.d.). Fixed fixture.

The United States Consumer Product Safety Commission. (2017). *2016 Annual Report to the President and Congress*.

U.S. Consumer Product Safety Commission. (1985). Consumer Product Safety Commission Annual Report.

U.S. Consumer Product Safety Commission. (n.d.). About us.

Wu, F., Samper, A., Morales, A. C., & Fitzsimons, G. J. (2017). It's too pretty to use! When and how enhanced product aesthetics discourage usage and lower consumption enjoyment. *The Journal of Consumer Research*, 44(3), 651–672. 10.1093/jcr/ucx057

Yim, S. (2020). Testing and certification industry in Hong Kong. HKTDC Research. https://research.hktdc.com/en/article/MzExMzYxMzAz

Chapter 4

Environmental Impact of Metal Additive Manufacturing

Mark Armstrong, H. Mehrabi, and Nida Naveed

CONTENTS

4.1 INTRODUCTION

The transition from the cottage industry, where goods were once crafted by hand until the emergence of mechanised factory systems in the late eighteenth to mid-nineteenth century, emerged as the prime mover for the first industrial revolution and a precursor for the second and more recent third. As the world stands on the threshold of the Fourth Industrial era, established factory and manufacturing systems are faced with a new upheaval. However, unlike previous industrial revolutions, the incentive for change does not exclusively arise from the influence of escalating consumerism and economies of scale; rather, the stimulus partly stems from the growing clamour to tackle climate change. There is now an increasing scientific consensus that the Earth is undergoing an unprecedented period of volatility brought about by the previous eras of industrialisation. Recently, nine processes and systems termed "planetary boundaries" have been identified that regulate the Earth's stability and resilience (Steffen et al., 2015). According to this model, infringing on one or more of these boundaries may bring about cataclysmic and non-linear environmental changes (Rockström

DOI: 10.1201/9781003306238-4

et al., 2009). So far, four of the nine boundaries have now been crossed, including climate change, altered biogeochemical cycles, land system change, and a loss of the biosphere.

Furthermore, the sixth assessment report of the Intergovernmental Panel on Climate Change, published in 2021 (IPCC, 2021), serves as a sobering reminder of humanity's role in bringing about the current climate crisis, warning that if human activity is left unregulated, within the next two decades, the Earth's temperature may rise by 1.5 °C and by up to 3 °C by the end of the century. For this reason, international efforts are now focused on mitigating the effects of climate change. Consequently, the agitation to tackle the global climate emergency has opened up debates around the sustainability of manufacturing processes. The 2030 agenda for sustainable development adopted by all United Nations members delivered a compilation of 17 intertwined goals designed to contend with such matters. In particular, goal number 12, responsible consumption and production (RCP), promotes vital changes in how societies produce and consume goods and services, an echo of the Oslo Symposium, which in part defined RCP as a way of reducing the consumption of natural resources throughout the lifecycle of a product by way of economically sound processes that minimise negative environmental impacts while conserving energy and natural resources. Although global industrial practices consume 15% to 40% of global energy and resources (Gao et al., 2021), the activities responsible for producing and processing metals are amongst the most energy-intensive (World Steel Association, 2021). Despite contributing more towards the economic development and enhancement of living standards than any other sector, the metal industry impacts land deprivation, the emission of GHGs, trace metals, particulate matter, and acidic gases, to name a few (Strezov et al., 2021). In Europe alone, the iron and steel industry in 2018 surpassed all other sectors emitting 22% of the total carbon dioxide (CO_2) emissions (European Parliament, 2020), along with the World Steel Association estimating that for every one ton of steel produced, the process emitted 1.85 tonnes of CO_2 (World steel, 2020). As the global population of roughly 7.8 billion people is projected to grow to 9.7 billion with an eventual peak in 2100 of nearly 11 billion, metal consumption will likely increase simultaneously, particularly as developing countries strive for comparable economic augmentation and as domestic material consumption levels become homogenised with other industrialised nations. While urgent and exceptional, present-day ecological challenges for the metal industry coincide with a period of unprecedented technological innovation, particularly with recent processing and metallurgy developments in MAM, which has the capability of printing many hard-to-process materials without expensive tooling, especially for intricate parts with internal fluid channels, such as heat exchangers, proton exchange membrane fuel cells, and artificial blood vessels, among others (Zhang et al., 2020). At the same time, the manufacturing workflow can be drastically condensed as other stages inherent to TM processes become redundant. For example, raw material needs to be melted, refined, and held for some time before casting in the foundry process, yet, in the MAM workflow, powder, filament, wire, and other feedstocks are typically acquired ready-made and can be immediately loaded into the machine ready for printing.

Furthermore, waste byproducts like metal chips from CNC machining may also be drastically reduced, as structures can be printed near net-shaped (NNS), meaning final products are geometrically as close as possible to the desired shape. However, extensive post-processing is often required to remove build plates, support structures and residual powder. Machining is also usually undertaken to improve the surface quality and tolerances, where additional hot isostatic pressing (HIP) is sometimes also carried out to reduce the material's porosity in addition to conventional heat treating methods; for example, performing solution annealing or ageing to adjust or customise the part's mechanical properties. Conversely, recent work has shown promise in developing novel solid-state processes that transform metal chips into useable and reliable powder (Jordon et al., 2020; Batista et al., 2021). Thus, considering many metals used are typically recyclable, MAM could play a vital role in the future by propagating circular economies by producing brand new products by reprocessing surplus material and supporting sustainable manufacturing. Despite the technology being in the relatively early stages of development, MAM displays potential economic and functional advantages over TM processes for some applications; however, whether or not it exhibits the same ecological benefits has yet to be adequately demonstrated. With an increasing number of companies now actively engaged with these novel technologies and many more considering their implementation, manufacturers should be able to evaluate the sustainable credentials of MAM to plan responsible manufacturing strategies in an era that is increasingly affected by climate change and resource scarcity. This chapter attempts to answer these questions by presenting an overview of the current state of knowledge on the environmental impact of MAM from the available literature. The first section of this chapter presents the most prominent and contemporary MAM technologies. The following section aims to give a transparent overview of the work that has already been done to characterise the environmental impact of MAM for the various technologies listed in Table 4.1. Based on the current literature, the objective is to compare each technology based on specific energy consumption (SEC) in kWh/kg. The results are then used to compare MAM with TM processes such as sand casting, CNC machining, forging, and powder metallurgy from the GRANTA CES database (Granta Design Limited (n.d.)). Lastly, recent developments in the domain of smart manufacturing are addressed in the context of SSAM, and the sustainability benefits of MAM are discussed.

Table 4.1 MAM processes per ISO/ASTM 52900:2017 modes and applications (Technical Committee AMT/8, 2017)

Process	Abbreviation	Synonyms	Mode	Typical Applications
Material extrusion	ME	FDM, FFF	Material is selectively dispensed through a nozzle or orifice.	Functional metal parts, prototyping.
Binder jetting	BJ	SPJ, NPJ, SD	A liquid bonding agent is selectively deposited to join powder materials.	Functional metal parts, low-rate production runs of non-critical components.
Directed energy deposition	DED	LMD, LENS, WAAM	Focused thermal energy fuses materials by melting as they are deposited.	Functional metal parts, repairs, adding material to existing parts.
Powder bed fusion	PBF	SLM, SLS, DMLS, DMLM	Thermal energy selectively fuses regions of a powder bed.	Functional metal parts, functional prototyping.

4.2 METAL ADDITIVE MANUFACTURING TECHNOLOGIES

MAM fabricates physical products from digital data by the successive addition of metal feedstock fused or bonded together in a layer-by-layer fashion (Technical Committee AMT/8, 2017). The number of different technologies is vast, as are the synonyms to describe each one, with each varying in its layer manufacturing method. Individual processes, therefore, differ depending on the material and system used. In 2010, rhe International Organisation for Standards (ISO) and American Society for Testing and Materials (ASTM) 52900:2017 were developed to establish methodologies and terminologies. The directive separates AM technologies into seven general categories; however, only the four most common MAM methods in Table 4.1 are considered for this chapter.

4.3 ENERGY CONSUMPTION AND CO$_2$ EMISSIONS

The characterisation of MAM technologies by their energy consumption has been investigated since the early 1990s. In order to quantify this, most types of analyses report this in terms of SEC, which is primarily used to understand and compare the energy consumption of TM processes (Yoon et al., 2014). Typically, studies express SEC as the energy used (in kWh or MJ) per kg of processed material. This chapter considers SEC data for the MAM printing and TM subtractive and formative processes, whereas the preprocessing and post-processing of material are not considered for MAM or TM methods. The total energy consumption can be regarded as comprising two parts for metal printing. First is the primary energy, representing the energy required to process the metal feedstock (e.g., powder and wire), which can also be regarded as direct printing energy. The final part is viewed as secondary energy, which comprises the energy required to power all auxiliary components of the machine (Peng, 2016). The total energy consumed in the entire printing process can be represented mathematically by Eq. (4.1):

$$E_{tot} = E_p + E_s \tag{4.1}$$

where E_p is the primary energy and E_s is the secondary energy. The SEC then, which is a ratio of E_{tot} over a given period of time divided by the sum of the final mass of the printed part, can be expressed by Eq. (4.2):

$$SEC = \frac{E_{tot}}{m_c} \tag{4.2}$$

where m_c is the mass in kg of the finished metal part. Typically, in energy measurement experiments for MAM, an electric meter or power data acquisition device, including voltage and current sensor and data acquisition cards, can be used to capture the energy consumed by the process (Majeed et al., 2020). In addition to collecting SEC from various journal articles, this chapter will also calculate how much gas is theoretically generated from the printing process, contributing to global warming relative to CO_2, typically known as CO_2 equivalent (CO_2-eq) in kg. Once SEC data has been identified, it is possible to estimate CO_2-eq emissions output using Eq. (4.3):

$$CO2_{eq} = SEC \times C_{int} \qquad (4.3)$$

where C_{int} is the carbon intensity. According to the UK Governments Greenhouse gas reporting: conversion factors, the average carbon intensity is 0.23314 kg CO_2/kWh (Hill et al., 2020).

4.4 DATA COLLECTION METHODOLOGY

The primary purpose of this chapter is to review and analyse the available literature based on previous work to understand the energy consumption of metal printing processes and compare them against TM methods. To achieve this goal, a comprehensive literature review on the SEC for the printing process for the technologies listed in Table 4.1 is performed through a qualitative systematic search strategy by defining several key synonyms. The prevailing emphasis is on the applicability of the search terms, summarising the evidence accurately and reliably, and uncovering all the relevant studies which address the following question:

- What work has been done to characterise the SEC for ME, BJ, PBF, and DED printing processes?

The entire qualitative search and subsequent filtering of results involved the following four stages:

- *Stage 1* was conducted using the EBSCO discovery service (EDS). The expressions 3D printing, additive manufacturing, and rapid prototyping have all been used to describe and generalise the technology. Therefore, searches included these synonyms to recognise all relevant literature. The search strings were limited to title, abstract and keywords to keep the number of articles controllable using a Boolean search strategy with the terms "AND" and "OR" to define relationships among synonyms. The term "metal" was also used; then, either "ME", "BJ", "PBF", or "DED" were used to focus the results. Table 4.2 documents all search terms conducted across the database,

Table 4.2 Stage 1 – Boolean search strings used for journal article discovery

ID	Boolean String Search Terms	Search Location	Limiters
A1	("additive manufacturing" OR "3D printing" OR "rapid prototyping") AND ("metal") AND ("energy" OR "energy consumption" OR "specific energy consumption" OR "power consumption")	Title, abstract, subject terms	English, 2012–2022
A2	("additive manufacturing" OR "3D printing" OR "rapid prototyping") AND ("metal") AND ("sustainability" OR "environmental")	Title, abstract, subject terms	English, 2012–2022
A3	("additive manufacturing" OR "3D printing" OR "rapid prototyping") AND ("metal") AND ("life cycle assessment" OR "life cycle analysis" OR "environmental assessment")	Title, abstract, subject terms	English, 2012–2022
B1	("material extrusion" OR "fused deposition modelling" OR "fused filament fabrication") AND ("metal") AND ("energy" OR "energy consumption" OR "specific energy consumption")	Title, abstract, subject terms	English, 2012–2022
B2	("material extrusion" OR "fused deposition modelling" OR "fused filament fabrication") AND ("metal") AND ("sustainability" OR "environmental")	Title, abstract, subject terms	English, 2012–2022
B3	("material extrusion" OR "fused deposition modelling" OR "fused filament fabrication") AND ("metal") AND ("life cycle assessment" OR "life cycle analysis" OR "environmental assessment")	Title, abstract, subject terms	English, 2012–2022
C1	("binder jetting") AND ("metal") AND ("energy" OR "energy consumption" OR "specific energy consumption")	Title, abstract, subject terms	English, 2012–2022
C2	("binder jetting") AND ("metal") AND ("sustainability" OR "environmental")	Title, abstract, subject terms	English, 2012–2022
C3	("binder jetting") AND ("metal") AND ("life cycle assessment" OR "life cycle analysis" OR "environmental assessment")	Title, abstract, subject terms	English, 2012–2022

(Continued)

Table 4.2 (Continued) Stage 1 – Boolean search strings used for journal article discovery

ID	Boolean String Search Terms	Search Location	Limiters
D1	("powder bed fusion" OR "selective laser melting" OR "selective laser sintering" OR "electron beam melting" OR "direct metal laser sintering") AND ("metal") AND ("energy" OR "energy consumption" OR "specific energy consumption")	Title, abstract, subject terms	English, 2012–2022
D2	("powder bed fusion" OR "selective laser melting" OR "selective laser sintering" OR "electron beam melting" OR "direct metal laser sintering") AND ("metal") AND ("sustainability" OR "environmental")	Title, abstract, subject terms	English, 2012–2022
D3	("powder bed fusion" OR "selective laser melting" OR "selective laser sintering" OR "electron beam melting" OR "direct metal laser sintering") AND ("metal") AND ("life cycle assessment" OR "life cycle analysis" OR "environmental assessment")	Title, abstract, subject terms	English, 2012–2022
E1	("directed energy deposition" OR "wire arc additive manufacturing" OR "laser metal deposition" OR "laser engineered net shaping") AND ("metal") AND ("energy" OR "energy consumption" OR "specific energy consumption")	Title, abstract, subject terms	English, 2012–2022
E2	("directed energy deposition" OR "wire arc additive manufacturing" OR "laser metal deposition" OR "laser engineered net shaping") AND ("metal") AND ("sustainability" OR "environmental")	Title, abstract, subject terms	English, 2012–2022
E3	("directed energy deposition" OR "wire arc additive manufacturing" OR "laser metal deposition" OR "laser engineered net shaping") AND ("metal") AND ("life cycle assessment" OR "life cycle analysis" OR "environmental assessment")	Title, abstract, subject terms	English, 2012–2022

Table 4.3 Stage 2 – Journal paper discovery from EDBS using terms from stage I

ID	Title	Abstract	Subject Terms	Total
A I	37	I478	I44	I659
A2	I0	I55	23	I88
A3	0	I6	3	I9
B I	0	36	I	37
B2	0	7	0	7
B3	0	0	0	0
C I	0	20	3	23
C2	I	3	I	5
C3	0	I	I	2
D I	I0	756	67	833
D2	I	82	I0	93
D3	0	8	2	I0
E I	43	519	76	638
E2	3	24	4	3I
E3	0	I	I	2
Total papers discovered from *stage I*				**35I3**

- *Stage 2* of the process utilised the 15 search terms from *Stage 1*, and 3,513 articles were uncovered, which are documented in Table 4.3,
- *Stage 3* uncovered 61 relevant articles in Table 4.4. Because of the vast number of papers, each abstract was briefly reviewed, and relevant articles were downloaded and stored according to their search term identity (e.g., A1 and A2). Mendeley Desktop application was used to organise the data that was collected,
- *Stage 4* utilised the "check for duplicates" function in the Mendeley Desktop application. After all duplicate studies were merged, 16 articles remained relevant to the study's purposes shown in Table 4.5.

Table 4.6 catalogues the details of each journal article from *stage 4*. The authors, publication date, MAM category and technology, the material used, the methodology, and the SEC were recorded. In addition, the papers were read in full, and any relevant references to other journal articles which had not been discovered in the initial search were also retrieved. Eight additional studies documented in Table 4.7 were recovered. In total, 29 studies that include identifiable SEC data have been found. The CES Granta database (Granta Design Limited (n.d.)) was then used to acquire the SEC for TM processes. The materials selected were mapped to the materials used for the MAM studies for continuity purposes. The SEC for these processes is shown in Table 4.8.

Table 4.4 Stage 3 – Filtering of irrelevant journal articles

ID	Title	Abstract	Subject Terms	Total
A1	2	12	1	
A2	0	16	2	18
A3	0	11	2	13
B1	0	0	0	0
B2	0	0	0	0
B3	0	0	0	0
C1	0	1	0	1
C2	1	1	1	3
C3	0	1	1	2
D1	1	7	0	8
D2	1	8	1	10
D3	0	3	2	5
E1	0	0	0	0
E2	0	0	1	1
E3	0	0	0	0
Total papers filtered from *stage 2*				61

Table 4.5 Stage 4 – Number of journal articles left after duplication removal

Journal articles remaining after removing duplicate studies	
Articles remaining after *stage 3*	16

4.5 SEC COMPARISON BETWEEN MAM AND TM

Of the 29 studies discovered, 17 were related to PBF, 4 for DED and 1 for BJ. Consequently, there is a considerable lack of understanding of some primary MAM technologies' energy consumption and environmental implications, especially for BJ and ME. Nevertheless, a comparison can be made with TM data from the Granta CES database and the data collected for PBF and DED. Figure 4.1 compares the SEC for these processes.

The comparison shows that MAM has a much higher SEC than most TM methods when only considering the additive, subtractive, or formative process. SEC for MAM processes vary significantly across the three categories, with a 2.47 kWh/kg lower limit attributable to the single BJ study. However, it is challenging to reliably compare this process with only one analysis relative to all others, and therefore it is important to establish the SEC of BJ with future studies. The range of SEC for DED is also considerable at 287.04 kWh/kg, showing that DED is the most energy-consuming process overall. Similarly, relative to PBF data, with only four journal articles dedicated to the energy consumption of DED, it would be helpful if

Table 4.6 SEC from relevant journal articles from search

ID	Reported by	Year	Ref.	Methodology	MAM Category	Mam Technology	Material	SEC (kWh/kg)	CO_2-eq (kg)	Ref. to other Studies (See Table 4.8)
S1	Jackson et al.	2016	(Jackson et al., 2016)	Comparative	DED	LENS	Carbon steel	16.97	3.96	(Morrow et al., 2007; Baumers et al., 2011, 2010; Kellens et al., 2011)
S2	Paris et al.	2016	(Paris et al., 2016)	Comparative LCA	PBF	EBM	Ti6AlV	104.2	24.29	...
S3	Bekker et al.	2016	(Bekker et al., 2016)	Experimental	DED	WAAM	SS	5.18	4.95	(Baumers et al., 2016)
S4	Nagarajan et al.	2017	(Nagarajan and Haapala, 2017)	LCA	PBF	DMLS	Iron	77.77	18.13	(Morrow et al., 2007)
S5	Nyamekye et al.	2017	(Nyamekye et al., 2017)	LCIA	PBF	PBF	316L SS	11.08	2.58	(Morrow et al., 2007; Baumers et al., 2011)
S6	Kellens et al.	2017	(Kellens et al., 2017a)	Review	...	...	...	...	...	(Kellens et al., 2011; Faludi et al., 2017; Baumers et al., 2017)
S7	Ingarao et al.	2018	(Ingarao et al., 2018)	LCA	...	...	...	...	...	(Baumers et al., 2011; Paris et al., 2016)
S8	Kamps et al.	2018	(Kamps et al., 2018)	LCA	PBF	LBM	16MnCr5	26.52	6.18	(Baumers et al., 2011, 2010; Kellens et al., 2011; Baumers et al., 2017, 2013)
S9	Liu et al.	2018	(Liu et al., 2018a)	Review	...	...	...	...	...	(Morrow et al., 2007; Baumers et al., 2011, 2010; Kellens et al., 2017a; Faludi et al., 2017; Wilson et al., 2014; Sachs et al., 1990)
S10	Nagarajan	2018	(Nagarajan and Haapala, 2018)	Exergy analysis	PBF	DMLS	Iron	155.17	36.18	(Morrow et al., 2007; Kellens et al., 2011)
S11a	Fruggiero et al.	2019	(Fruggiero et al., 2019)	Comparative analysis	PBF	DMLS	Ti-6Al-4V	62.22	14.51	(Morrow et al., 2007; Kellens et al., 2011; Paris et al., 2016; Kellens et al., 2017a; Faludi et al., 2017)

(Continued)

Table 4.6 (Continued) SEC from relevant journal articles from search

ID	Reported by	Year	Ref.	Methodology	MAM Category	Mam Technology	Material	SEC (kWh/kg)	CO₂-eq (kg)	Ref. to other Studies (See Table 4.8)
S11b	Fruggiero et al.	2019	(Fruggiero et al., 2019)	Comparative analysis	PBF	DMLS	Ti-6Al-4V	31.11	7.25	(Morrow et al., 2007; Kellens et al., 2011; Paris et al., 2016; Kellens et al., 2017a; Faludi et al., 2017)
S11c	Fruggiero et al.	2019	(Fruggiero et al., 2019)	Comparative analysis	PBF	DMLS	AlSi10Mg	93.89	21.89	(Morrow et al., 2007; Kellens et al., 2011; Paris et al., 2016; Kellens et al., 2017a; Faludi et al., 2017)
S11d	Fruggiero et al.	2019	(Fruggiero et al., 2019)	Comparative analysis	PBF	DMLS	PH1	63.33	14.76	(Morrow et al., 2007; Kellens et al., 2011; Paris et al., 2016; Kellens et al., 2017a; Faludi et al., 2017)
S12	Ahmad et al.	2020	(Ahmad and Enemuoh, 2020)	LCA	PBF	EBM	316L SS	21.05	4.91	(Baumers et al., 2011, 2010; Kellens et al., 2011; Paris et al., 2016)
S13	Majeed et al.	2020	(Majeed et al., 2020)	Review	...	...	...	...	...	(Morrow et al., 2007; Baumers et al., 2011, 2010; Kellens et al., 2011; Wilson et al., 2014; Le Bourhis et al., 2014)
S14	Peng et al.	2020	(Peng et al., 2020)	LCA	PBF	SLM	316L SS	19.79	4.61	(Morrow et al., 2007; Paris et al., 2016)
S15	Raoufi et al.	2020	(Raoufi et al., 2020)	LCA	BJ	BJ	316L SS	2.47	0.58	...
S16	Monteiro et al.	2022	(Monteiro et al., 2022)	Review	...	...	...	...	...	(Paris et al., 2016; Ingarao et al., 2018)

Table 4.7 SEC from journal articles referenced by studies in the initial search

Study ID	Reported by	Year	Ref.	Methodology	MAM Category	MAM Technology	Material	SEC (kWh/kg)	CO_2-eq (kg)
S20	Sachs et al.	1990	(Sachs et al., 1990)	Experimental	PBF	SLM	316L SS	26.89	6.27
S21a	Baumers et al.	2010	(Baumers et al., 2010)	Comparative + experimental	PBF	SLM	316L SS	31.00	7.23
S21b	Baumers et al.	2010	(Baumers et al., 2010)	Comparative + experimental	PBF	EBM	Ti-6Al-4V	17.00	3.96
S21a	Baumers et al.	2011	(Baumers et al., 2011)	Experimental	PBF	EBM	Ti-6Al-4V	16.94	3.95
S21b	Baumers et al.	2011	(Baumers et al., 2011)	Experimental	PBF	EBM	Ti-6Al-4V	49.17	11.46
S21c	Baumers et al.	2011	(Baumers et al., 2011)	Experimental	PBF	SLM	316L SS	23.05	5.37
S21d	Baumers et al.	2011	(Baumers et al., 2011)	Experimental	PBF	SLM	316L SS	29.44	6.86
S21e	Baumers et al.	2011	(Baumers et al., 2011)	Experimental	PBF	DMLS	17-4 PH SS	66.94	15.61
S21f	Baumers et al.	2011	(Baumers et al., 2011)	Experimental	PBF	DMLS	17-4 PH SS	94.17	21.95
S21g	Baumers et al.	2011	(Baumers et al., 2011)	Experimental	PBF	SLM	316L SS	117.5	27.39
S21h	Baumers et al.	2011	(Baumers et al., 2011)	Experimental	PBF	SLM	316L SS	163.33	38.08
S22	Kellens et al.	2011	(Kellens et al., 2011)	LCIA	PBF	SLM	316L SS	26.94	6.28
S23	Baumers et al.	2013	(Baumers et al., 2013)	Estimation tool	PBF	LBM	17-4 PH SS	77.78	18.13
S24	Wilson et al.	2014	(Wilson et al., 2014)	Experimental	DED	LDD	Nistelle 625	292.22	68.13
S25	Bourhis et al.	2014	(Le Bourhis et al., 2014)	Predictive model	DED	LENS	Metal (not specified)	24.2	5.64
S26a	Baumers et al.	2016	(Baumers et al., 2016)	Experimental	PBF	EBM	Ti-6Al-4V	38.40	8.95
S27b	Baumers et al.	2016	(Baumers et al., 2016)	Experimental	PBF	DMLS	17-4 PH SS	62.90	14.66
S28	Faludi et al.	2016	(Faludi et al., 2017)	LCA	PBF	SLM	AlSi10Mg	157.28	36.67
S29	Baumers et al.	2016	(Baumers et al., 2017)	Experimental	PBF	EBM	Ti-6Al 4V	16.67	3.89

Table 4.8 SEC for TM processes from CES Granta database

Reported by	TM Process	Material	SEC (kWh/kg)	CO₂-eq (kg)	Ref.
CES Granta database	Casting	17-4 PH SS	3–3.31	0.70–0.77	(Granta Design Limited (n.d.))
	Coarse machining	17-4 PH SS	0.48–0.55	0.11–0.13	
	Fine machining	17-4 PH SS	4.31–4.75	1.00–1.11	
	Powder metallurgy	17-4 PH SS	10.28–11.28	2.40–2.63	
	Forging	17-4 PH SS	2.86–23.17	0.67–5.40	
	Casting	316L SS	1.08–1.2	0.25–0.28	
	Coarse machining	316L SS	0.21–0.23	0.05–0.05	
	Fine machining	316L SS	0.89–0.99	0.21–0.23	
	Powder metallurgy	316L SS	10.28–11.28	2.40–2.63	
	Forging	316L SS	0.59–0.65	0.14–0.15	
	Casting	Ti-6Al-4V	3.86–4.23	0.90–0.99	
	Coarse machining	Ti-6Al-4V	0.20–0.70	0.05–0.16	
	Fine machining	Ti-6Al-4V	5.14–5.67	1.20–1.32	
	Powder metallurgy	Ti-6Al-4V	12.67–14.08	2.95–3.28	
	Casting	AlSi10mg	2.94–3.25	0.69–0.76	
	Coarse machining	AlSi10mg	0.29–0.33	0.07–0.08	
	Fine machining	AlSi10mg	1.75–1.94	0.41–0.45	

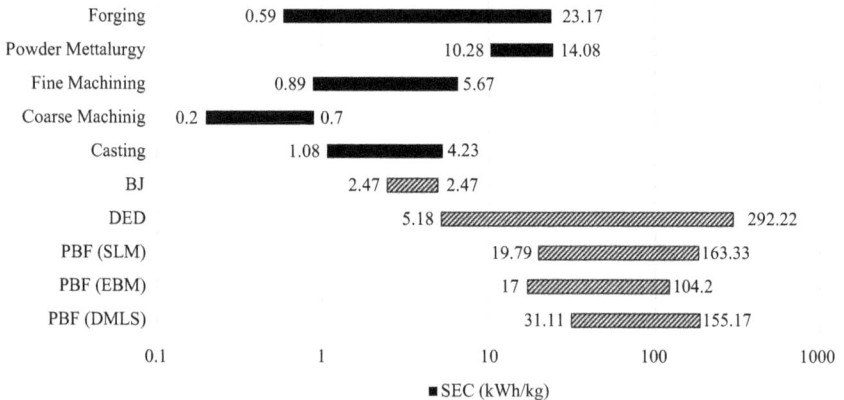

Figure 4.1 Specific energy consumption comparison of MAM and TM processes.

more contemporary work was done in this area, especially as the information was established in 2014 and 2016. For PBF, SEC is relatively similar across all sub-processes, with a 146.33 kWh/kg range overall. SLM has a range of 143.54 kWh/kg, EMB has a range of 87.2 kWh/kg, and DMLS has a range of 124.06 kWh/kg; therefore, the EBM process is the most energy-efficient PBF technology. Although relatively lower than most MAM methods, the SEC for TM processes also fluctuates considerably. The lowest energy-consuming process is coarse machining at 0.5 kWh/kg, whereas the most energy-consuming, with the most extensive SEC range, is forging at 22.58 kWh/kg.

4.5.1 SEC comparison based on material type

To elaborate on the variation in SEC for both TM and MAM processes and based on the information at hand, the data can be arranged to compare SEC concerning the processed material. Figure 4.2 shows the SEC for processing 17-4 PH stainless steel. The data for this material is derived from three journal articles. The most energy-intensive method is DMLS, and the energy range for forging is relatively substantial at 20.31 kWh/kg.

Figure 4.3 shows the SEC for processing 316L stainless steel. The data illustrates that the SLM process is the most energy-intensive method overall. Data for SLM originates from eight journal articles, whereas the data for EBM is derived from only one study. The data range for SLM is wide-ranging at 143.54 kWh/kg.

Figure 4.4 shows the comparison for processing Ti-6Al-4V. Of the two PBF processes, DMLS is the most energy-efficient, with a 31.11 kWh/kg range compared to SLM at 87.53 kWh/kg.

Figure 4.5 shows the SEC for processing AlSi10Mg. Although the data for DMLS and SLM are the result of only two separate studies, the SEC for both MAM processes is significantly higher than the SEC for TM processes.

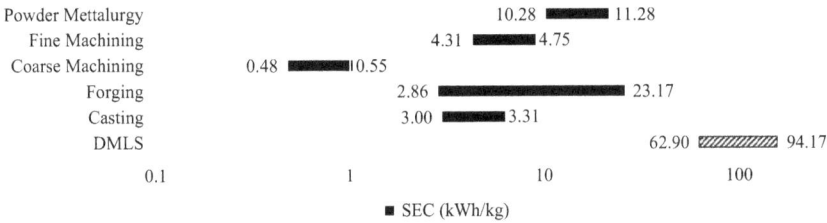

Figure 4.2 Specific energy consumption comparison for 17-4 PH stainless steel.

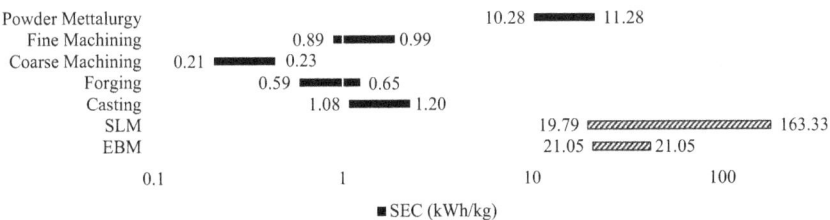

Figure 4.3 Specific energy consumption comparison for 316L stainless steel.

Powder Mettalurgy   12.67 ▬▬ 14.08
Fine Machining   5.14 ▬▬ 5.67
Coarse Machining   0.20 ▬▬▬▬▬ 0.70
Casting   3.86 ▬▬ 4.23
DMLS   31.11 ▨▨▨▨ 62.22
EBM   16.67 ▨▨▨▨▨▨▨104.20
                    0.1   1   10   100
                              ■ SEC (kWh/kg)

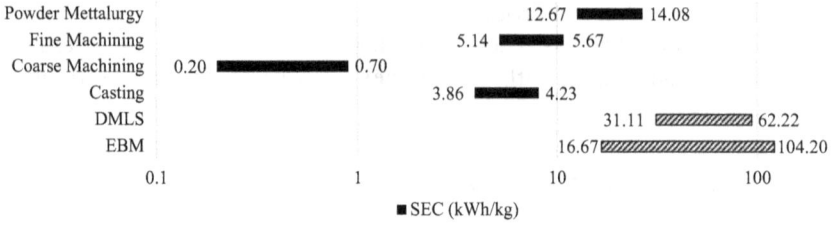

Figure 4.4 Specific energy consumption comparison for Ti-6Al-4V.

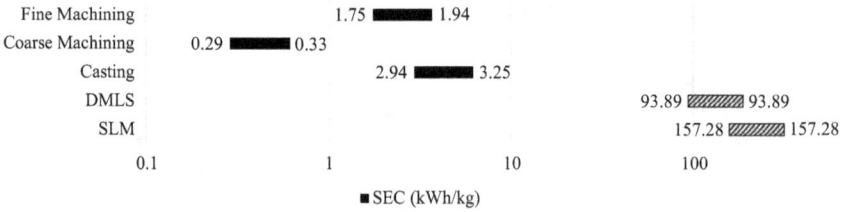

Fine Machining   1.75 ▬▬ 1.94
Coarse Machining   0.29 ▬▬▬ 0.33
Casting   2.94 ▬▬ 3.25
DMLS   93.89 ▨▨ 93.89
SLM   157.28 ▨▨ 157.28
                    0.1   1   10   100
                              ■ SEC (kWh/kg)

Figure 4.5 Specific energy consumption comparison for AlSi10Mg.

4.5.2 Limitations of SEC for MAM printing

The manufacturing process (e.g., printing, forming, or subtracting) is only one process in the product's entire life cycle. To conclude with certainty, the actual environmental impact of a manufacturing process, the whole life cycle of a product needs to be quantified, including material extraction, primary material processing, secondary material processing, post-processing, and disposal. In the absence of complete life cycle assessments, the data collected for this chapter suggests that the energy demand of MAM is typically greater than most TM processes, making MAM the less sustainable manufacturing option overall. Forming the same verdict for technologies such as DED and BJ is challenging due to the relative lack of comparative data, thus making the environmental implications unclear. The impact of ME is even more ambiguous due to a total lack of scrutiny on the sustainable characteristics of the technology in the scientific literature. Several reviews also substantiate the conclusions of this comparison in terms of the energy demand of the MAM process, such as the work done by Kellen et al. (2017b), who demonstrated that the SEC for some AM processes are one, and, in some cases, two-fold higher than those for TM processes and Yoon et al. (2014) who also show that the SEC for some additive processes to be roughly one-fold higher than traditional subtractive processes. Although the accumulative printing energy of MAM is typically higher than TM methods, it could be claimed that the SEC found in the literature is simply an energy indicator due to complex input process variables (e.g., part orientation, packing density, laser power and scanning speed), material density, material properties, layer thickness, surface finish, and geometry which may also rationalise the large variations of SEC. As already mentioned, SEC is typically composed of both primary and secondary energy where in addition to the energy required to melt or sinter the feedstock, auxiliary equipment reliant on secondary energy such as heaters, pumps, chillers, and controllers may also vary considerably between machines and thus may also influence the overall SEC which may also account for the expansive range in SEC for some MAM processes. Comparing MAM to TM is complex as the implications of only comparing energy consumption per kg are unclear, as the functional unit (FU) can be manufactured in very different ways and in various shapes and sizes. For example, with MAM, the material is added; for CNC machining, the material is incrementally removed from the billet, and for forging and casting, the material is normally formed to the shape of the desired product. Consequently, depending on the complexity of geometry, SEC may vary significantly. Typically, in energy experiments, the geometry of the FU is a 1 kg dense cube; however, (Van Sice and Faludi 2020), in a similar study, point out that in order to compensate for the disparity between manufacturing methods, a set of FUs designed to represent a standardised reference for both AM and TM processes, similar to the examples proposed by Priarone et al. (2017), should be adopted for comparative energy consumption experiments. Van Sice and Faludi (2020) also recommend that at least one of the standard parts in the set represent components that would characteristically be created with AM machines, such as complex parts that include internal channels or are hollow, which would be challenging to manufacture with TM methods to allow for a fair assessment. In addition, most research into energy consumption has been performed in academic settings instead of manufacturing environments, where experimental procedures vary from study to study, which makes comparing and interpreting SEC data even more challenging. Despite measuring the impact per kg of printed material across all studies, there has been the need to develop quantitative methods to evaluate MAM processes equally and by the same standards and benchmarks. The review by Khalid and Peng (2021) detailed these very issues and found that several attempts have been made to develop reliable energy auditing frameworks to account for the variations of MAM processes such as the IoT and data analytics techniques that rely on methods such as linear regression, decision tree and back-propagation neural networks. As it stands, there is no single accepted criterion or standard protocol for researchers to follow when conducting energy experiments (Rejeski et al., 2018). However, with increased attention to process optimisation through in situ monitoring and advanced systems like BDA and ML, enough data may eventually be collected to characterise the relationship between process parameters and energy consumption. For instance, Liu et al. (2018b) successfully demonstrated the direct relationship between input parameters and the energy efficiency of the LENS process. By that token, and with enough data that characterises the process variations and their correlation to energy consumption, various models could be developed to enable the development of intelligent MAM systems that can autonomously adjust and optimise multiple process parameters in situ to reduce the energy consumption of the process.

4.6 THE ROLE OF SMART MANUFACTURING FOR SUSTAINABLE MAM

The notion of intelligent MAM systems or smart manufacturing (SM) suggested in the previous section refers to a holistic approach to improving and optimising all aspects of the manufacturing process through the synergy of various digital and physical technologies. Although process optimisation is already well established in industries like the aerospace sector, which implemented lean six sigma methodologies in the 1990s and 2000s to improve processes, Industry 4.0 (I4.0) creates new opportunities to harness the potential of digital technologies to deliver process improvements through the ability to continuously monitor, maintain, adjust and improve the functioning of manufacturing systems. Conceived initially in Germany, I4.0 aims to create advanced and intelligent factories by upgrading manufacturing systems to smart cyber-physical systems (CPSs) with IoT and cloud computing integration and with minimal human intervention (Wang et al., 2021). In these types of unified technologies, manufacturing systems will be able to monitor the manufacturing process in real-time using various sensory equipment to capture vast amounts of data that can be interpreted to make autonomous decisions based on various techniques such as BDA and ML. Consequently, the prospect of influencing the sustainability of manufacturing processes by I4.0 methodologies has recently gained increasing attention. However, not much attention has been focused on the impact of I4.0 on energy reduction (Haddouche et al., 2022); although several studies (Mohamed et al., 2019) have suggested that CPSs may significantly improve energy efficiency, this has not been sufficiently demonstrated (Ghobakhloo and Fathi, 2021).

On the other hand, the concept of SSAM proposed by Majeed et al. (2021), which relates to the consolidation of several disciplines converging to optimise the process parameters needed to improve product quality, reduce energy consumption, and improve productivity, could help to improve the energy efficiency of MAM printing processes. At the core of SSAM is the application of BDA, which can be leveraged to uncover and recognise patterns to identify relationships during the printing process. These patterns can then be used in computer programmes to optimise the design or acted upon in situ to adjust the process parameters. For instance, Majeed et al. (2019) demonstrated that by optimising the process parameters within a BDA framework in conjunction with a data-miningscheme, energy consumption could be reduced, and product quality and production efficiency could be improved by minimising processing time. Similarly, Qin et al. (2017) applied an IoT framework to the SLS process by harvesting multisource data from the process, which was analysed by BDA tools and then uploaded to the cloud to establish energy consumption, whereby a control system was proposed to optimise energy consumption. While SSAM machines with improved energy efficiency are yet to be developed, there is great potential for these systems in the coming years in a bid to enhance the sustainability of the manufacturing landscape. Without demonstratable results in the area of additive technology, energy savings have been described by businesses that have attempted to implement I4.0 for other technologies. For instance, the ISA reported that by configuring their robotic systems with I4.0 techniques, German company Daimler recorded a 30% improvement in energy efficiency, and by using sensory equipment to alert that energy consumption was outside of normalparameters, Canadian Forest Products was able to reduce their energy consumption by 15% (International Society for Automation (n.d.)).

4.7 SUSTAINABLE BENEFITS OF MAM

Despite the relatively high energy consumption of the printing process, several environmental advantages over TM processes are typically recognised in the literature.The first and most cited benefit is a reduction of materials and resources (Despeisse and Ford, 2015). Considering the notion of efficiency in manufacturing terms, the concept's ideology usually refers to the relationship between product output and resource input, or, in other words, achieving more with less. Insubtractive and formative manufacturing, up to 95% of the material can be wasted (Girdwood et al., 2017). For instance, metal chips that are the byproduct of the CNC machining process, which are removed to define the product's shape, are typically thrown away, and in casting, the material used to create supplementary runners and risers is also discarded after the part has solidified. In contrast, MAM only deposits material integral to producing NNS products making the process 97% theoretically efficient, though this figure is often less in reality (Peng et al., 2018). Some MAM technologies also often require supplementary material to create structures to support overhanging features,whichare disposed of at the post-processing stage.

Unlike surplus metal chips that are typically difficult and costly to recycle due to inherent machining oils and lubricants, up to 95% of unmelted powder in laser powder-based processes can be directly reused (Arrizubieta et al., 2020) following a recovery, recycling and reconditioning process (Daraban et al., 2019). However, some studies show that powders can deteriorate after several cycles (Powell et al., 2020; Raza et al., 2021; Kong et al., 2021), and research is limited to many commercially available metals and alloys in circulation today. An example of how doing more with less through the implementation of MAM to improve efficiency is in the aerospace sector, where the buy-to-fly ratio, which is a measure of the ratio of the weight of billet material purchased to the weight of the material of the finalpart, is typically in the range of 10:1–20:1 from using parts manufactured by TM processes (Blakey-Milner et al., 2021; Yusuf et al., 2019) meaning that for every 1 kg of material used in the aircraft, roughly 20 kg is discarded (Lockett, 2019) (i.e., 95% material is scrapped and only 5% is utilised). Conversely, the implementation of MAM enables a buy-to-fly ratio of almost 1:1, meaning material utilisation is approximately 95%, whereas only 5% is wasted due to the layer-by-layer process of creating NNS parts (Yusuf et al., 2019).

Similarly, the approach to how MAM parts are designed may not only improve the overall efficiency of the process but also reduce the environmental impact of the products themselves by applying frameworks such as design for additive manufacturing (DfAM) which may include methodologies such as topology optimisation and latticing additions (Plocher and Panesar, 2019). By optimising the distribution of material within the confines of the product specification for certain loads and boundary conditions while satisfying the required structural performance of a product, the areas of the part that are not supporting the applied loads and not undergoing substantial deformation and thus not contributing to the overall integrity of the component can be removed. Although optimisation strategies can be applied to TM, topologically optimised structures are typically characterised by complex geometric shapesthat are difficult and nearly impossible to produce by conventional means. In contrast, MAM removes these limitations with its layer-by-layer approach. Therefore, not only is less material required which also shortens the build time meaning less energy is needed for the printing process, the mass of the product can be significantly reduced, which is especially relevant for sustainable manufacturing. For example, in the aerospace industry, Gao et al. (2021) report that for every 1 kg of material removed, 90,000 litres of fuel could be saved, avoiding the release of roughly 230 tonnes of CO_2 into the atmosphere. Similarly, Kellens et al. (2017b) report on studies that estimate that using additively manufactured lightweight components in aircraft between 2019–2050 could save between 1.2 and 2.8 billion GJin energy, thereby preventing the release of 92.1–215.0 million tonnes of GHGs. Aside from the environmental benefits of efficient manufacturing, the entire supply chain can also be simplified compared to conventional and complex global supply chains. Therefore, the MAM process can be considered a driver for decentralised manufacturing systems, reducing transport and material flows with fewer manufacturing steps, fewer suppliers, and thus fewer environmental impacts. Given the advantages of MAM in the above examples, it is clear that the sustainability benefits emerge not as a result of the process itself but as a function of the capabilities of the technology based on its ability to manufacture complex geometries. How important a part MAM will play in the future of sustainable manufacturing is still unclear due to its relatively short existence; however, the potential for MAM to contribute to sustainable manufacturing practices is perceptible given its fundamental capabilities.

4.8 CONCLUSION

The present chapter presented a comprehensive review of the energy consumption of MAM by investigating what work has been done to characterise the environmental impact of the printing process based on SEC by performing a systematic literature review. The results were then compared to the SEC of TM processes such as casting, CNC machining, forging, and powder metallurgy. At the manufacturing process level, the results of this comparison indicate that MAM demands significantly more energy per kg than TM, with a range of 289.75 kWh/kg, whereas the range for combined TM processes was 22.97 kWh/kg. Results were then organised and compared depending on the material classification in an attempt to account for the significant range of SEC of the MAM process; however, any such relationships are unclear. The limitations and implications of only characterising the environmental impact of MAM based on SEC were discussed, along with the need to develop standardised methodologies to objectively measure the MAM process's energy consumption against TM methods. The significance of SM and I4.0 was then discussed as a conceivable way of reducing the energy consumption of additive processes in the future. Although the impact of advanced CPSs has yet to be demonstrated, their potential appears promising. Finally, the existing advantages of MAM for sustainable manufacturing were investigated, with the most apparent benefit being the consolidation of the manufacturing life cycle and to be able to optimise the design of products thanks to MAM's ability to manufacture complex shapes and, therefore, lighter products, which are challenging to produce by TM methods. Consequently, by adopting DfAM methodologies to print only geometry integral to the product's overall performance, resources, and waste can be significantly reduced, and overall efficiency and sustainability can be improved.

REFERENCES

N. Ahmad, E.U. Enemuoh, Energy modeling and eco impact evaluation in direct metal laser sintering hybrid milling, *Heliyon*. 6 (2020) e03168. 10.1016/j.heliyon.2020. e03168

J.I. Arrizubieta, O. Ukar, M. Ostolaza, A. Mugica, Study of the environmental implications of using metal powder in additive manufacturing and its handling, *Metals (Basel)*. 10 (2020) 261. 10.3390/met10020261

C.D. Batista, A.A.M. das N. de P. Fernandes, M.T.F. Vieira, O. Emadinia, From machining chips to raw material for powder metallurgy—A review, *Materials (Basel)*. 14 (2021) 5432. 10.3390/ma14185432

M. Baumers, P. Dickens, C. Tuck, R. Hague, The cost of additive manufacturing: Machine productivity, economies of scale and technology-push, *Technol. Forecast. Soc. Change.* 102 (2016) 193–201. 10.1016/j.techfore.2015.02.015

M. Baumers, C. Tuck, R. Hague, I. Ashcroft, R. Wildman, A comparative study of metallic additive manufacturing power consumption, in: 21st Annu. Int. Solid Free. Fabr. Symp. – An Addit. Manuf. Conf. SFF 2010, 2010: pp. 278–288.

M. Baumers, C. Tuck, R. Wildman, I. Ashcroft, R. Hague, Energy inputs to additive manufacturing: Does capacity utilisation matter?, in: 22nd Annu. Int. Solid Free. Fabr. Symp. – An Addit. Manuf. Conf. SFF 2011, 2011: pp. 30–40.

M. Baumers, C. Tuck, R. Wildman, I. Ashcroft, R. Hague, Shape complexity and process energy consumption in electron beam melting: A case of something for nothing in additive manufacturing?, *J. Ind. Ecol.* 21 (2017) S157–S167. 10.1111/jiec.12397

M. Baumers, C. Tuck, R. Wildman, I. Ashcroft, E. Rosamond, R. Hague, Transparency built-in: Energy consumption and cost estimation for additive manufacturing, *J. Ind. Ecol.* 17 (2013) 418–431. 10.1111/j.1530-9290.2012.00512.x

A.C.M. Bekker, J.C. Verlinden, G. Galimberti, Challenges in assessing the sustainability of wire + arc additive manufacturing for large structures, in: Solid Free. Fabr. 2016 Proc. 27th Annu. Int. Solid Free. Fabr. Symp. – An Addit. Manuf. Conf. SFF 2016, 2016: pp. 406–416.

B. Blakey-Milner, P. Gradl, G. Snedden, M. Brooks, J. Pitot, E. Lopez, M. Leary, F. Berto, A. du Plessis, Metal additive manufacturing in aerospace: A review, *Mater. Des.* 209 (2021) 110008. 10.1016/j.matdes.2021.110008

D. Cheng Kong, C. Fang Dong, X. Qing Ni, L. Zhang, R. Xue Li, X. He, C. Man, X. Gang Li, Microstructure and mechanical properties of nickel-based superalloy fabricated by laser powder-bed fusion using recycled powders, *Int. J. Miner. Metall. Mater.* 28 (2021) 266–278. 10.1007/s12613-020-2147-4

A.E.O. Daraban, C.S. Negrea, F.G.P. Artimon, D. Angelescu, G. Popan, S.I. Gheorghe, M. Gheorghe, A deep look at metal additive manufacturing recycling and use tools for sustainability performance, *Sustain.* 11 (2019) 5494. 10.3390/su11195494

M. Despeisse, S. Ford, The role of additive manufacturing in improving resource efficiency and sustainability, in: *IFIP Adv. Inf. Commun. Technol.*, Springer, Cham, 2015: pp. 129–136. 10.1007/978-3-319-22759-7_15

European Parliament, Energy-intensive industries challenges and opportunities in energy transition policy department for economic, *Scientific and Quality of Life Policies Directorate-General for Internal Policies*, (2020).

J. Faludi, M. Baumers, I. Maskery, R. Hague, Environmental impacts of selective laser melting: Do printer, powder, or power dominate?, *J. Ind. Ecol.* 21 (2017) S144–S156. 10.1111/jiec.12528

F. Fruggiero, A. Lambiase, R. Bonito, M. Fera, The load of sustainability for additive manufacturing processes, *Procedia Manuf.* 41 (2019) 375–382. 10.1016/j.promfg.2019.09.022

C. Gao, S. Wolff, S. Wang, Eco-friendly additive manufacturing of metals: Energy efficiency and life cycle analysis, *J. Manuf. Syst.* 60 (2021) 459–472. 10.1016/j.jmsy.2021.06.011

M. Ghobakhloo, M. Fathi, Industry 4.0 and opportunities for energy sustainability, *J. Clean. Prod.* 295 (2021) 126427. 10.1016/j.jclepro.2021.126427

R. Girdwood, M. Bezuidenhout, P. Hugo, P. Conradie, G. Oosthuizen, D. Dimitrov, Investigating components affecting the resource efficiency of incorporating metal additive manufacturing in process chains, *Procedia Manuf.* 8 (2017) 52–58. 10.1016/j.promfg.2017.02.006

Granta Design Limited, CES EduPack software, (n.d.).

M. Haddouche, A. Ilinca, Energy efficiency and Industry 4.0 in wood industry: A review and comparison to other industries, *Energies.* 15 (2022) 2384. 10. 3390/EN15072384

N. Hill, R. Bramwell, E. Karagianni, L. Jones, J. MacCarthy, S. Hinton, C. Walker, B. Harris, Government greenhouse gas conversion factors for company reporting: Methodology paper, Dep. Business, *Energy Ind. Strateg.* 2020 (2020) 128.

G. Ingarao, P.C. Priarone, Y. Deng, D. Paraskevas, Environmental modelling of aluminium based components manufacturing routes: Additive manufacturing versus machining versus forming, *J. Clean. Prod.* 176 (2018) 261–275. 10.1016/ j.jclepro.2017.12.115

International Society for Automation, Combining IoT, Industry 4.0, and energy management, (n.d.). https://www.isa.org/intech-home/2018/march-april/ features/combining-iot-industry-4-0-and-energy-management (accessed March 31, 2022).

IPCC, Sixth Assessment Report, (2021). https://www.ipcc.ch/assessment-report/ar6/ (accessed February 19, 2022).

M.A. Jackson, A. Van Asten, J.D. Morrow, S. Min, F.E. Pfefferkorn, A comparison of energy consumption in wire-based and powder-based additive-subtractive manufacturing, in: *Procedia Manuf.*, Elsevier, 2016: pp. 989–1005. 10.1016/ j.promfg.2016.08.087

J.B. Jordon, P.G. Allison, B.J. Phillips, D.Z. Avery, R.P. Kinser, L.N. Brewer, C. Cox, K. Doherty, Direct recycling of machine chips through a novel solid-state additive manufacturing process, *Mater. Des.* 193 (2020) 108850. 10.1016/ j.matdes.2020.108850

T. Kamps, M. Lutter-Guenther, C. Seidel, T. Gutowski, G. Reinhart, Cost- and energy-efficient manufacture of gears by laser beam melting, *CIRP J. Manuf. Sci. Technol.* 21 (2018) 47–60. 10.1016/j.cirpj.2018.01.002

K. Kellens, E. Yasa, Renaldi, W. Dewulf, J.P. Kruth, J.R. Duflou, Energy and resource efficiency of SLS/SLM processes, 22nd Annu. Int. Solid Free. Fabr. Symp. – An Addit. Manuf. Conf. SFF 2011. (2011) 1–16.

K. Kellens, R. Mertens, D. Paraskevas, W. Dewulf, J.R. Duflou, Environmental impact of additive manufacturing processes: Does AM contribute to a more sustainable way of part manufacturing?, in: *Procedia CIRP*, Elsevier, 2017a: pp. 582–587. 10.1016/j.procir.2016.11.153

K. Kellens, M. Baumers, T.G. Gutowski, W. Flanagan, R. Lifset, J.R. Duflou, Environmental dimensions of additive manufacturing: Mapping application domains and their environmental implications, *J. Ind. Ecol.* 21 (2017b) S49–S68. 10.1111/jiec.12629

M. Khalid, Q. Peng, Sustainability and environmental impact of additive manufacturing: A literature review, *Comput. Aided. Des. Appl.* 18 (2021) 1210–1232. 10.14733/cadaps.2021.1210-1232

F. Le Bourhis, O. Kerbrat, L. Dembinski, J.Y. Hascoet, P. Mognol, Predictive model for environmental assessment in additive manufacturing process, in: *Procedia CIRP*, Elsevier, 2014: pp. 26–31. 10.1016/j.procir.2014.06.031

Z. Liu, Q. Jiang, F. Ning, H. Kim, W. Cong, C. Xu, H.C. Zhang, Investigation of energy requirements and environmental performance for additive manufacturing processes, *Sustain*. 10 (2018b) 3606. 10.3390/su10103606

Z.Y. Liu, C. Li, X.Y. Fang, Y.B. Guo, Energy consumption in additive manufacturing of metal parts, *Procedia Manuf*. 26 (2018a) 834–845. 10.1016/j.promfg.2018.07.104

H. Lockett, Additive Manufacturing – Helping to reduce waste in aircraft production, *Des. Res*. (2019). http://www.open.ac.uk/blogs/design/additive-manufacturing-helping-to-reduce-waste-in-aircraft-production/ (accessed March 22, 2022).

A. Majeed, A. Ahmed, J. Lv, T. Peng, M. Muzamil, A state-of-the-art review on energy consumption and quality characteristics in metal additive manufacturing processes, *J. Brazilian Soc. Mech. Sci. Eng*. 42 (2020). 10.1007/s40430-020-02323-4

A. Majeed, J. Lv, T. Peng, A framework for big data driven process analysis and optimisation for additive manufacturing, *Rapid Prototyp. J*. 25 (2019) 308–321. 10.1108/RPJ-04-2017-0075

A. Majeed, Y. Zhang, S. Ren, J. Lv, T. Peng, S. Waqar, E. Yin, A big data-driven framework for sustainable and smart additive manufacturing, *Robot. Comput. Integr. Manuf*. 67 (2021) 102026. 10.1016/j.rcim.2020.102026

N. Mohamed, J. Al-Jaroodi, S. Lazarova-Molnar, Industry 4.0: Opportunities for enhancing energy efficiency in smart factories, in: SysCon 2019 – 13th Annu. IEEE Int. Syst. Conf. Proc., Institute of Electrical and Electronics Engineers Inc., 2019. 10.1109/SYSCON.2019.8836751

H. Monteiro, G. Carmona-Aparicio, I. Lei, M. Despeisse, Energy and material efficiency strategies enabled by metal additive manufacturing – A review for the aeronautic and aerospace sectors, *Energy Reports*. 8 (2022) 298–305. 10.1016/j.egyr.2022.01.035

W.R. Morrow, H. Qi, I. Kim, J. Mazumder, S.J. Skerlos, Environmental aspects of laser-based and conventional tool and die manufacturing, *J. Clean. Prod*. 15 (2007) 932–943. 10.1016/j.jclepro.2005.11.030

H.P.N. Nagarajan, K.R. Haapala, Environmental performance evaluation of direct metal laser sintering through exergy analysis, *Procedia Manuf*. 10 (2017) 957–967. 10.1016/j.promfg.2017.07.087

H.P.N. Nagarajan, K.R. Haapala, Characterising the influence of resource-energy-exergy factors on the environmental performance of additive manufacturing systems, *J. Manuf. Syst*. 48 (2018) 87–96. 10.1016/j.jmsy.2018.06.005

P. Nyamekye, H. Piili, M. Leino, A. Salminen, Preliminary investigation on life cycle inventory of powder bed fusion of stainless steel, in: *Phys. Procedia*, Elsevier, 2017: pp. 108–121. 10.1016/j.phpro.2017.08.017

H. Paris, H. Mokhtarian, E. Coatanéa, M. Museau, I.F. Ituarte, Comparative environmental impacts of additive and subtractive manufacturing technologies, *CIRP Ann. – Manuf. Technol*. 65 (2016) 29–32. 10.1016/j.cirp.2016.04.036

T. Peng, Analysis of energy utilization in 3D printing processes, in: *Procedia CIRP*, Elsevier B.V., 2016: pp. 62–67. 10.1016/j.procir.2016.01.055

T. Peng, K. Kellens, R. Tang, C. Chen, G. Chen, Sustainability of additive manufacturing: An overview on its energy demand and environmental impact,
Addit. Manuf. 21 (2018) 694–704. 10.1016/j.addma.2018.04.022

T. Peng, Y. Wang, Y. Zhu, Y. Yang, Y. Yang, R. Tang, Life cycle assessment of
selective-laser-melting-produced hydraulic valve body with integrated design
and manufacturing optimisation: A cradle-to-gate study, *Addit. Manuf.* 36
(2020) 101530. 10.1016/j.addma.2020.101530

J. Plocher, A. Panesar, Review on design and structural optimisation in additive
manufacturing: Towards next-generation lightweight structures, *Mater. Des.*
183 (2019) 108164. 10.1016/j.matdes.2019.108164

D. Powell, A.E.W. Rennie, L. Geekie, N. Burns, Understanding powder degradation
in metal additive manufacturing to allow the upcycling of recycled powders,
J. Clean. Prod. 268 (2020) 122077. 10.1016/j.jclepro.2020.122077

P.C. Priarone, G. Ingarao, R. di Lorenzo, L. Settineri, Influence of material-related
aspects of additive and subtractive Ti-6Al-4V manufacturing on energy
demand and carbon dioxide emissions, *J. Ind. Ecol.* 21 (2017) S191–S202.
10.1111/jiec.12523

J. Qin, Y. Liu, R. Grosvenor, A framework of energy consumption modelling for
additive manufacturing using internet of things, in: *Procedia CIRP*, Elsevier
B.V., 2017: pp. 307–312. 10.1016/j.procir.2017.02.036

K. Raoufi, S. Manoharan, T. Etheridge, B.K. Paul, K.R. Haapala, Cost and environmental impact assessment of stainless steel microreactor plates using
binder jetting and metal injection molding processes, *Procedia Manuf.* 48
(2020) 311–319. 10.1016/j.promfg.2020.05.052

A. Raza, T. Fiegl, I. Hanif, A. MarkstrÖm, M. Franke, C. Körner, E. Hryha,
Degradation of AlSi10Mg powder during laser based powder bed fusion processing, *Mater. Des.* 198 (2021). 10.1016/j.matdes.2020.109358

D. Rejeski, F. Zhao, Y. Huang, Research needs and recommendations on environmental implications of additive manufacturing, *Addit. Manuf.* 19 (2018)
21–28. 10.1016/j.addma.2017.10.019

J. Rockström, W. Steffen, K. Noone, Å. Persson, F.S. Chapin, E. Lambin, T.M.
Lenton, M. Scheffer, C. Folke, H.J. Schellnhuber, B. Nykvist, C.A. de Wit,
T. Hughes, S. van der Leeuw, H. Rodhe, S. Sörlin, P.K. Snyder, R. Costanza,
U. Svedin, M. Falkenmark, L. Karlberg, R.W. Corell, V.J. Fabry, J. Hansen,
B. Walker, D. Liverman, K. Richardson, P. Crutzen, J. Foley, Planetary
boundaries: Exploring the safe operating space for humanity, *Ecol. Soc.* 14
(2009). 10.5751/ES-03180-140232

E. Sachs, M. Cima, J. Cornie, Three-dimensional printing: Rapid tooling and prototypes directly from a CAD model, *CIRP Ann. – Manuf. Technol.* 39 (1990)
201–204. 10.1016/S0007-8506(07)61035-X

J. Van Sice, C. Faludi, State of knowledge on the environmental impacts of metal
additive manufacturing, (2020).

Van Sice, Corrie, & Faludi, Jeremy (2021). COMPARING ENVIRONMENTAL
IMPACTS OF METAL ADDITIVE MANUFACTURING TO CONVENT-
IONAL MANUFACTURING. *Proceedings of the Design Society*, 1, 671–
68010.1017/pds.2021.67.

W. Steffen, K. Richardson, J. Rockström, S.E. Cornell, I. Fetzer, E.M. Bennett, R.
Biggs, S.R. Carpenter, W. De Vries, C.A. De Wit, C. Folke, D. Gerten, J.
Heinke, G.M. Mace, L.M. Persson, V. Ramanathan, B. Reyers, S. Sörlin,

Planetary boundaries: Guiding human development on a changing planet, *Science (80-.).* 347 (2015). 10.1126/science.1259855

V. Strezov, X. Zhou, T.J. Evans, Life cycle impact assessment of metal production industries in Australia, *Sci. Rep.* 11 (2021) 10116. 10.1038/s41598-021-89567-9

Technical Committee AMT/8, BS EN ISO/ASTM 52900: Additive mnufacturing – General principles – Terminology, *Int. Stand.* (2017) 1–30.

B. Wang, F. Tao, X. Fang, C. Liu, Y. Liu, T. Freiheit, Smart manufacturing and intelligent manufacturing: A comparative review, *Engineering.* 7 (2021) 738–757. 10.1016/j.eng.2020.07.017

World steel, Steel's Contribution to a low carbon future – worldsteel position paper, (2020). https://www.steel.org.au/resources/elibrary/resource-items/steel-s-contribution-to-a-low-carbon-future-and-cl/ (accessed August 4, 2021).

World Steel Association, Climate change and the production of iron and steel, World Steel Assoc. (2021) 8. https://www.steel.org.au/resources/elibrary/resource-items/steel-s-contribution-to-a-low-carbon-future-and-cl/ (accessed August 7, 2021).

J.M. Wilson, C. Piya, Y.C. Shin, F. Zhao, K. Ramani, Remanufacturing of turbine blades by laser direct deposition with its energy and environmental impact analysis, *J. Clean. Prod.* 80 (2014) 170–178. 10.1016/j.jclepro.2014.05.084

H.S. Yoon, J.Y. Lee, H.S. Kim, M.S. Kim, E.S. Kim, Y.J. Shin, W.S. Chu, S.H. Ahn, A comparison of energy consumption in bulk forming, subtractive, and additive processes: Review and case study, *Int. J. Precis. Eng. Manuf. - Green Technol.* 1 (2014) 261–279. 10.1007/s40684-014-0033-0

S.M. Yusuf, S. Cutler, N. Gao, Review: The impact of metal additive manufacturing on the aerospace industry, *Metals (Basel).* 9 (2019) 1286. 10.3390/met9121286

C. Zhang, S. Wang, J. Li, Y. Zhu, T. Peng, H. Yang, Additive manufacturing of products with functional fluid channels: A review, *Addit. Manuf.* 36 (2020) 101490. 10.1016/j.addma.2020.101490

Chapter 5

Identification and Overcoming Key Challenges in the 3D Printing Revolution

Ashish Kaushik, Upender Punia, Sumit Gahletia,
Ramesh Kumar Garg, and Deepak Chhabra

CONTENTS

DOI: 10.1201/9781003306238-5

5.1 INTRODUCTION

The rise of three-dimensional printing (3DP) techniques has paved the way for creating profoundly complex geometry and inside structures, attributable to its extraordinary layer-wise assembling approach. This rapidly evolving technology permits the user to make personalized products resulting in cost reduction by eliminating the additional expense of tools and molds during the fabrication. Moreover, additive manufacturing (AM) empowers the development of intricate and consolidated utilitarian designs in a single-step operation, thus diminishing the requirement for assembling the complete part. The AM processes hold incredible potential for working on material productivity and decreasing life cycle environmental impacts compared to conventional techniques, which assists with advancing a sustainable "green" and economical manufacturing strategy. Three-dimensional printing is the fabrication method involving a machine driven by a computer, incrementally adding the layers of material to create a three-dimensional structure (Sharma et al., 2022).

The technology has the enormous potential to produce significant changes to the economy, global logistics, and the environment (Campbell et al., 2011). Prototype fabrication has always been one of the vital steps in product development, which can be time-intensive depending upon the complexity of objects. Although 3D printing has been generally used for fabricating prototypes for past decades, it is now approaching the point of inflection as a popular, serial manufacturing process resulting in effective product development, manufacturing performance, and enhanced product performance. AM also supports the computationally guided design approach specifically for shape optimization and the creation of various materials surpassing the prevailing benchmarks. New product development is presently the most significant variable driving firm achievement or disappointment for numerous businesses. To be effective for a new product, a company should amplify the fit with client needs and limit the time to market.

The seeds for 3D printing current industrial progress were initially planted in the 1980s through the invention of numerous technologies and the gradual adoption of an additive manufacturing process for rapid prototyping across manufacturing industries (Quinlan and Hart, 2020). Pioneers of these technologies shaped their ideas into companies by commercializing such as 3D Systems (Stereolithography) and Stratasys (Polymer-based extrusion), which now lead to considerable market share enclosed by industry. The parts fabricated by AM processes in their infancy phase are generally fragile and coarse; advancements in hardware, software, and materials have equipped additive manufacturing for mainstream adoption. The involvement of industrial stakeholders and participants have been exponentially increased due to the expiration of various fundamental patents in the past few decades. The firm's attention is focused on digitally guided production models and businesses that perform more effectively requiring less human labor, infrastructure, and other resources for producing a flexible catalogue of parts in response to shifting consumer preferences and supply chain uncertainty.

AM can reduce or eliminate the requirement of product or part-specific and fabricate intricate geometries that integrate material parts, higher material efficiency level, and combine materials in earlier impossible ways. Therefore, AM can considerably affect the capital costs, costs of flexibility, and marginal production costs (Koren, 2006). Previous research concluded that economic attributes are so much distinct that investment decisiveness into AM proves to be highly strategic (Li et al., 2014). Additive manufacturing processes influence the market structure apart from direct impacts on an individual firm's production methods. A substantial increase in the "makers" community is noticed who generate and share 3D models, fabricate 3D printers for household usage, and sell 3D printed parts (Schneider, 2006). The use of 3DP for fabricating prototypes is a trivial method for countless manufacturers. It allows manufacturers and suppliers to bypass the costly, laborious, and expensive conventional processes, for instance, the production of molds, casts and dies, milling, and several other machines perform the task and shipping the object from a supplier eventually. Three-dimensional printing can accelerate the new product development cycle, which can get new products to market frequently and at a quicker pace, especially in the case of intricate parts.

A considerable increase in the use of 3D printers in industries and home usage further stretches the opportunities for manufacturing processes. There is an indication that this disruptive technology is on the apex of being mainstreamed. Hence, there are insights into the possible opportunities and disruptions that rapid prototyping could create, including:

1. Industries investing in 3D printing see a remarkable change in speed, quick launching of new products, flexibility, and product customization.
2. Primitive adopters of this technology are crossing the threshold limits from beginner and prototype to the manufacturer of the end product.
3. Companies started to imagine the supply chains from a different perspective that can be described as the globe of networked 3D printers where the logistics are about bringing digitized design files from the continent to continents instead of containers, cargo, ships, and planes. A survey led by PwC innovations indicates that almost 70% of manufacturers firmly believe that 3D printing can be used to print obsolete objects, and 57% of them found the useful role of 3d printing in fabricating after-market products. In comparison, 30% of respondents are firmly in agreement about the 3D printing techniques greatest agitation will be exerted on the supply chains.

4. Industries and manufacturing companies are expecting 3DP-guided savings in labor, material along with transportation costs in contrast with conventional subtractive manufacturing techniques. A PwC investigation of 3D printing adoption by the aerospace industry's maintenance, repair, and operations (MRO) parts market accounts for a $3.4 billion saving in material and transportation costs annually (Wohlers, 2013; Dumitrescu and Tanase, 2016; Magisetty and Cheekuramelli, 2019).

5.2 CURRENT AND FUTURE CHALLENGES ARISING IN 3D PRINTING ADOPTION

AM is the latest trend and gaining traction in production processes due to its outstanding benefits (Abdulhameed et al., 2019). While this disruptive technology has already begun to appear in certain sectors, individual businesses are still struggling to accept and use it on a commercial basis. Various AM sectors such as materials, service, hardware, and software are foreseen to progress rapidly in the coming years (Davies, 2020). Companies overlook how much rapid 3D printing adoption is a learning process that spans the entire value chain and necessitates exceptional AM AM can decrease the demand for the manufacturing chain due to factors like near net shape character during the creation of smart factories (Deloitte, 2017). A blockage to sophisticated technology, structural organisation, and the accompanying ecosystem in which manufacturing companies operate are just a few of the significant hurdles that may hinder the growth and success of implementation during this learning process. Figure 5.1 gives an overview of pertinent

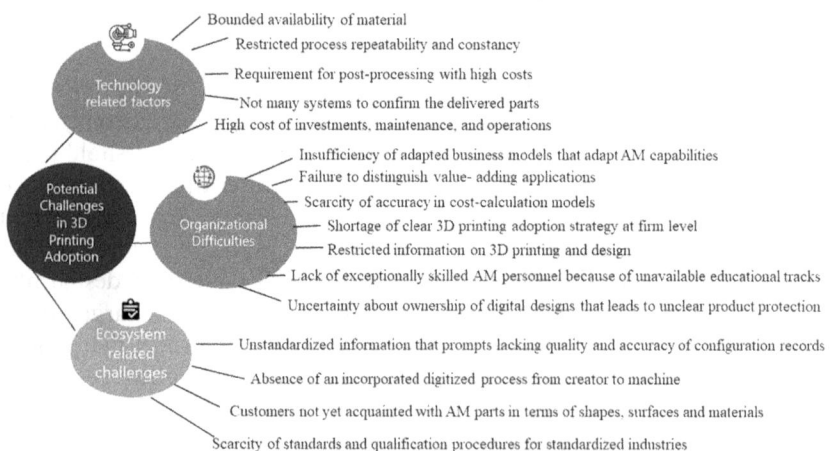

Figure 5.1 Potential challenges in 3D printing adoption and implementation.

challenges arising in the adoption and implementation of AM techniques (Omidvarkarjan et al., 2022; Basso et al., 2022).

The manufacturing costs of additively manufactured parts are a major obstacle during the adoption of AM. Experts indicate that material and machine costs, along with the dearth of quality and robustness have a considerable influence on costs. These factors are collectively prompt to exalt quality impudence after rapid prototyping processes. Intricate geometrical configurations of additively manufactured parts, small batch size, etc., result in the difficulty to post-process and mechanized manufacturing.

Scarcity of specified standards in regulated areas like rail, aerospace, automotive, or health sector slows up the rise of the 3D printing market. Standardization is obstruction, execution of AM technology demands industries to simplify ambiguity, safeguard liability and pass the approval process. However, the attempts from several organizations such as International Organization for Standardization (ISO) and American Society for Testing and Materials (ASTM) have considerably contributed to AM standards. International accredited registrar and various classified societies like Det Norske Veritas-Germanischer Lloyd (DNV-GL) are continuously collaborating with research and technology organizations and industrial parts for generating metal AM guidelines (AMFG, 2018). Therefore, quality production and certification of additively manufactured parts are viewed as a fundamental challenge for the upcoming decade by immense industrial experts surveyed.

Three-dimensional printing is also characterized simply as a digitized process. Individual digital production circumstances, including mass customization, demand an exceptionally proficient digital backbone. Although disruption of the digital process chain is an ongoing issue that is expected to be resolved over time, the quality is degraded by the connectivity of software from one provider to the next, as well as the interconnectedness of the process chain and intervening procedures. Moreover, the information or data created along the process chain is frequently accessed or shared.

For acceptable utilization and successful execution of 3D printing in their firm, a high degree of expertise is required. For example, today's cost-competitive goods must make use of 3D printing's advantages. This issue can be fixed through constant advancement in skills and knowledge in the industry and by students getting into the workforce subsequently following a module of 3D printing processes in the upcoming years.

5.3 CURRENT OPPORTUNITIES ENABLED BY 3D PRINTING

Although the aforesaid roadblocks and challenges appear to be disheartening at first instance, numerous booming implementations of 3D printing in real-world parts and the comprehensive opportunities developed by AM hugely override its drawbacks (Ukobitz, 2020). Generally, 3D printing

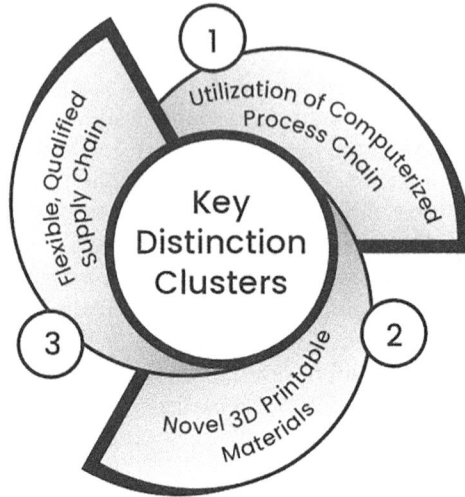

Figure 5.2 Key distinction clusters for advanced 3D printing utilizations.

consists of a variety of processes that have the potential to translate virtual data into physical objects (Huang et al., 2015). Three fundamental clusters of highly modern 3D printing utilizations are discussed in this section. Figure 5.2 showcase the fundamental distinction of collectively; they empower firms to develop new ways of capturing value, for instance, utilization of novel and digital business models.

5.3.1 Utilization of the computerized process chain

AM technology is digital in nature and compatible with huge varieties of other prevailing technology from Industry 4.0 for consideration. The underlying difference between the planning of additively manufacturing process and conventional subtractive technique lies in AM requirement of various software resulting in incompatibilities arising inflow of information across various steps (Denkena et al., 2018). Artificial intelligence, data-guided designs, and supply chains are all areas of focus. This unparalleled ability is mainstreamed by the digital thread that stretches the AM process chain. Data acquisition can be done to enhance the efficiency and expenditure of 3D printing applications throughout the entire product life cycle.

The following study by 3D systems (3DSystems) and Point designs (Point designs) demonstrates the way a computerized process chain can be controlled to develop exceptionally advanced 3D printing applications. Simultaneously, the firms create a mechanized design platform for the printing of partial hand prosthetic components. The corresponding design platform includes a variety of automation levels like simple parameter manipulation in addition to more complex features like the automatic fitting of 3D-printed hand prosthetics to the person's 3D scan. The solution is

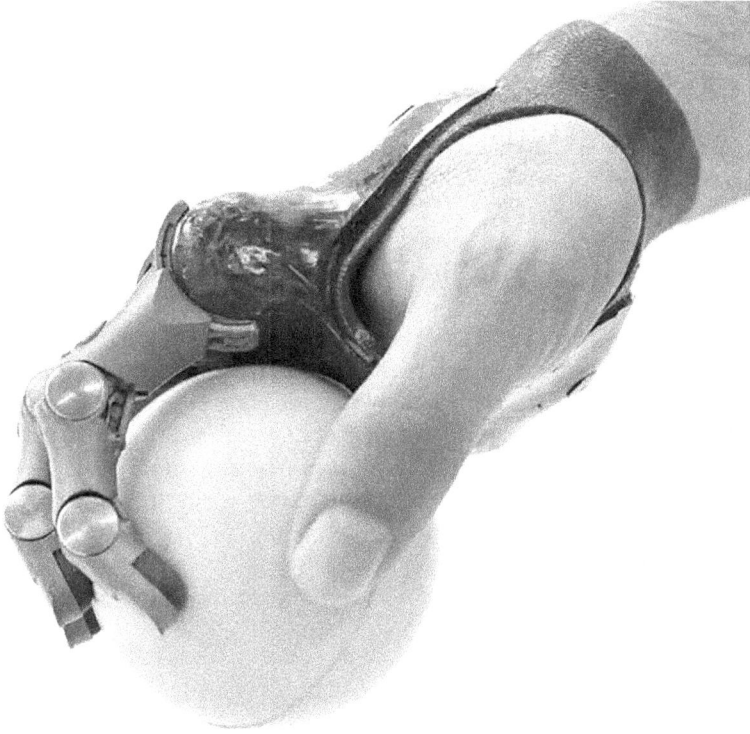

Figure 5.3 Automatically designed 3D printed partial hand prosthetic component.

achieved by collaborating to advance design for 3D printing, consulting the optimized printing strategy for mass production, followed by the selection of appropriate material.

Figure 5.3 demonstrates the automatically designed orthosis by 3D printing. This leads to enhanced customer interaction. The above example clearly illustrates the role of the digital process chain in AM processes by automation of particular process steps. This results in a wide variety of novel opportunities in context with the selling of products and customer interaction from a business perspective.

5.3.2 Novel 3D printable materials

In the era of Industry 4.0, creating novel AM material has been a long-term standing matter of interest for researchers and industry experts. Novel printing techniques and materials support rapid and multi-material printing, resulting in new opportunities and applications (Layani et al., 2018). The need arises because of the frugal range of novel additively manufactured materials, which proves to be a major barrier to adopting and implementing AM. Development of various novel materials has opened up the way for several new AM applications along with increased performance and decreased cost.

A study by Johnson and Jhonson (J&J) (Johnson & Johnson., 2020) clearly demonstrates the use of novel materials for AM utilization. The company brings forward a patient-specific implant for treating several clinical conditions. It consists of a personalized cage acting as structural support during bone healing issues. Depending upon the computed tomography (CT) scan, a personalized design is developed in a quick span of time. Following the surgeon's approval, the appropriate component is made in a central location and quickly delivered to the place of treatment. Figure 5.4 illustrates the fabrication of personalized implants for bone defects.

The highly intricate geometrical structures are made using the laser sintering technology and a polycaprolactone-based material blend (PCL). Because of its bioresorbable nature, it is progressively absorbed by the body over time (Punia et al., 2022). To improve the incorporation of the implant into the surrounding bone structure, calcium phosphate is then coated on the implant. The integration of novel AM materials with a particular design

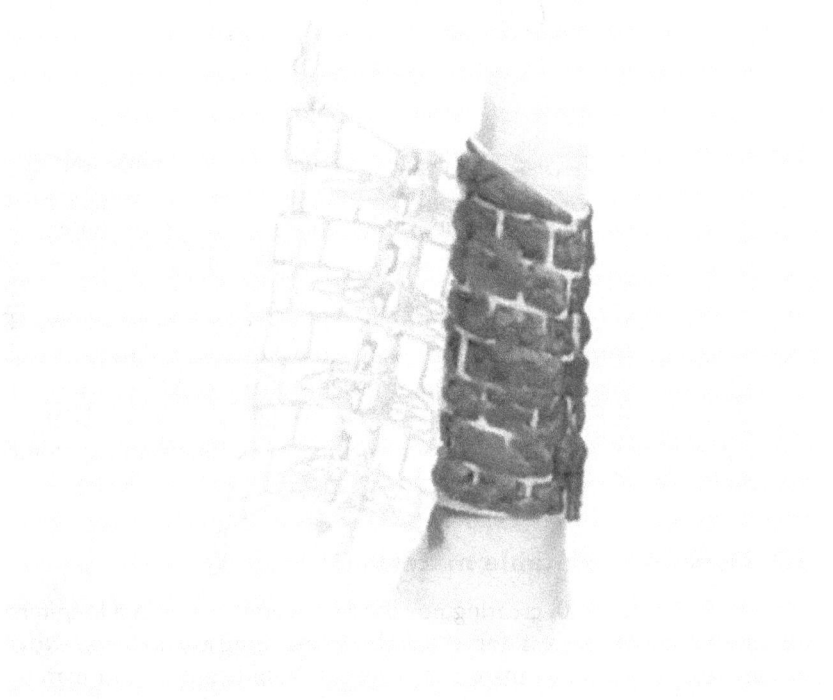

Figure 5.4 Personalized 3d printed implant for crucial-sized segmental bone defects.

followed by specific post-processing multiplies the benefits in contrast with existing solutions. These consist of faster bone remodelling, higher bine volume, and improved torsional strength. The appropriate combination of manufacturing techniques, novel materials, and digital process chain results in effective functionality and availability at a competitive price.

5.3.3 Flexible, qualified supply chain networks

Aside from the promise of creative freedom, 3D printing's unequalled supply chain has paved the path for businesses to use this cutting-edge technology. Three-dimensional printing processes have spawned a rapidly developing business, as well as the emergence of new service providers. This includes the ability to make components in tiny batch quantities without dramatically increasing prices, as well as decentralised manufacturing close to the point of application. Three-dimensional printing has the potential to establish supply chains that are more flexible and resilient than traditional procedures, allowing them to step in when traditional methods fail.

The following study by Deutsche Bahn (DB) (Deutsche Bahn) and TÜV SÜD (TÜV SÜD) demonstrates how to build a strong AM supply chain. DB has a comparably extensive fleet of various freight and passenger vehicles. Because of the infrastructure and the fact that these vehicles were constructed by different manufacturers throughout the years, maintaining them and ensuring spare parts availability is a difficult undertaking. AM proves to be an alternative in this context as it allows economical and rapid prototyping of individual, hard-to-source parts. However, the barrier lies in strict regulations of the railways for produced AM parts. DB collaborated with TUVSUD to create a certification scheme for suppliers dealing with additively manufactured spare parts and end-use products. The initiation of a general certification scheme and quality assurance standard leads to manifold benefits. The report above emphasizes the necessity of AM process standardization as a requirement for robust and flexible supply chains. Figure 5.5 illustrates the motor mount fabricated spare parts with 3D printing techniques.

5.4 ADOPTION STRATEGIES AND BEST PRACTICES FOR 3D PRINTING

The visionary framework for implementing additive manufacturing coming into the picture was put forward and is demonstrated in Figure 5.2. The conceptual framework offered that both internal strategy and external forces guide the concern of 3D printing as a process of manufacture, and the corresponding approach in the implementation of 3D printing will be influenced by factors which can be classified as four practical recommendations as represented in Figure 5.6 (Mellor et al., 2014).

Figure 5.5 Motor mount fabricated as spare part with 3D printing.

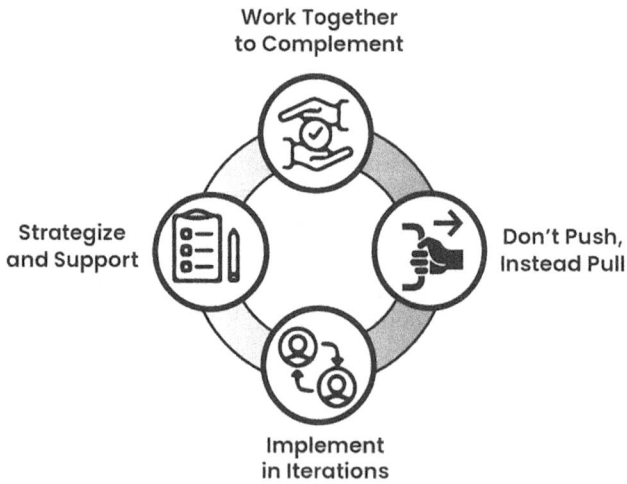

Figure 5.6 Best practices for 3D printing adoption.

5.4.1 Iterate the process

Several research studies concluded that the execution of AM techniques requires a wide variety of adequacy coming into the picture due to its cost-effective approach (Norrish et al., 2021). Three-dimensional printing adoption can be started with the learning process in which organizations gain AM expertise via verified training (Martinsuo and Luomaranta, 2018). The elevated hype of learning can be supervised by implementing an iterative and validated learning mode. Industries or corporations may progress to more complex applications by gradually increasing their AM capability and adoption.

5.4.2 Pull instead push

Generally, the adoption of AM is a costly capital-intensive procedure due to lofty investment in research and education (Sonar et al., 2020). To increase the likelihood of many successful implementations, to develop a self-sustaining adoption environment, 3D printing techniques should be backed by compelling client needs and a good business case. In contrast to a technology push strategy, in which structural organizations display their technical capabilities in unnecessarily complicated agitators, this market-pull method is quite different. Rapid prototyping methodologies face a number of challenges, including technological feasibility, cost-benefit analysis, and identifying an effective business model.

5.4.3 Complement each other's efforts

As previously stated, the utilization of AM technologies necessitates a wide range of knowledge across the entire 3D printing process chain. It can be challenging for some companies to cover all of the components internally, especially at the start of the adoption process. Such businesses benefit from collaborating with external collaborators, such as contract manufacturers, research institutions, or service providers, to close gaps in capability or competence.

5.4.4 Strategize and support

Several studies underline the importance that execution of AM techniques requires strategic significance in the adopting firm as long as important changes in operational structure are needed. Being at the intersection of invention, technology, and operations management, a choice is taken in favor of AM methods for industrial component manufacture (Oettmeier and Hofmann, 2017). To tackle this problem, companies are emboldened to define a 3D printing strategy ahead of time and align explicit execution tasks with this plan. For effective implementation, chief executive officers (CEOs) must give higher-level management assistance in the form of a cash budget and personal commitment.

5.5 KEY ENABLER AND SOLUTION APPROACH

From the several studies and challenges, here are the six most vital enablers for AM industrialization, along with possible solutions:

5.5.1 Cost savings in manufacturing and the process chain

A considerable escalation in technical breakthrough techniques such as binder jetting and process speed in research have showcased a higher potential to attenuate the manufacturing cost in the coming decade. In addition, material cost is further expected to decrease because of the novel or improved material production systems and new scale economics. Mechanization of the process chain is still in its early stages of development, and it may result in additional cost assets. It is expected that 3D printing methods will get a better understanding of design-to-cost.

5.5.2 Production qualification and quality confirmation

In today's scenario, it is possible to fabricate excellent quality and intricate 3D printed parts. However, a less intricate and expensive qualification strategy for rapid prototyping is required to empower forthcoming production. The machine characteristics, technical details or specifications of quality impudence, and 3D printing materials are categorized in mutual coordination for this purpose. In a recent study, intensification of quality impudence has been a prominent emphasis. In addition, the development of simplified quality impudence and qualification is expected to be moderate and stable.

5.5.3 Consolidation of digital process chain and 3D printing into manufacturing

In the current context, industries that invest with a comprehensive strategic plan and have consistent management assistance have the ability to entirely establish AM as a manufacturing technology. A computerized chain of processes, such as software for operation planning, has the capability to vastly boost productivity. Experts firmly believe that additive manufacturing processes can be perceived as consolidated manufacturing technology with ripened digital process chain.

5.5.4 Establishment of additional standards and guidelines

Standardization in 3D printing is progressing rapidly. However, the 3D printing process, data formats, digital security, and sustainability assessment demand outstanding standardization. The broad use of consistent standards and norms is expected to improve the overall quality of 3D printing, as well as its acceptability and industrialization across all sectors.

5.5.5 New application development

The effectiveness of the 3D printing market ruler is contingent on reevaluating procedures and applications with an entrepreneurial attitude. Choosing the right purpose and design for 3D printing is predicted to be a

Requirements for 3D Printing
Forthcoming Trends

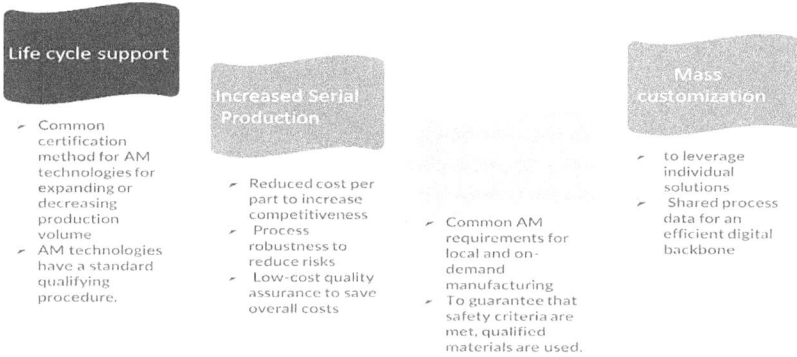

Life cycle support

- Common certification method for AM technologies for expanding or decreasing production volume
- AM technologies have a standard qualifying procedure.

Increased Serial Production

- Reduced cost per part to increase competitiveness
- Process robustness to reduce risks
- Low-cost quality assurance to save overall costs

- Common AM requirements for local and on-demand manufacturing
- To guarantee that safety criteria are met, qualified materials are used.

Mass customization

- to leverage individual solutions
- Shared process data for an efficient digital backbone

Figure 5.7 Requirements for 3D printing forthcoming trends.

significant facilitator in business cases. Several studies show that the usage of automated software for component design has increased. Researchers are also looking at new 3D printed materials and combinations, as well as multi-material solutions, to expand the range of applications. Current advancements are anticipating a new business case, as well as a market expansion for specialist applications.

5.5.6 Expertise development

The expansion and development of 3D printing expertise across the sector, particularly among small and medium-sized businesses (SMEs), may help to propel the total market forward. Experts believe that advancement will accelerate significantly in the next years, but they still see it as a vital facilitator for the next decade. The progressive increase of skills in 3D printing enterprises necessitates a long-term effort. Specific requirements for 3D printing future production trends can be obtained with the help of six sigma enablers which are illustrated in Figure 5.7.

5.6 KEY ENABLERS NECESSITATE SPECIFIC ACTIONS

To foster the accomplishment of key enablers, this section concludes with a necessitate specific actions for the worldwide production community to bring 3D printing industrialization a step ahead. All the stakeholders in the 3D printing ecosystem require putting efforts to conquer the issues built by 3D printing technique commercialization intricacy and to provide the 3D printing capability to a massive scale in the industries. This section discusses

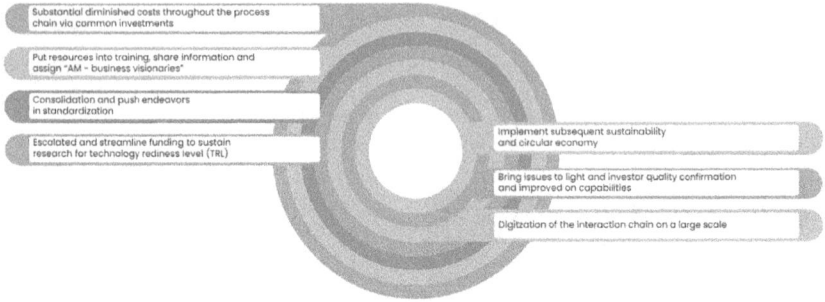

Substantial diminished costs throughout the process chain via common investments

Put resources into training, share information and assign "AM – business visionaries"

Consolidation and push endeavors in standardization

Escalated and streamline funding to sustain research for technology rediness level (TRL)

Implement subsequent sustainability and circular economy

Bring issues to light and investor quality confirmation and improved on capabilities

Digitzation of the interaction chain on a large scale

Figure 5.8 Significant steps to advance 3D printing.

various steps used for the advancement of additive manufacturing processes, as demonstrated in Figure 5.8.

5.6.1 Substantial diminished costs throughout the process chain via common investments

Researchers and technology providers are being encouraged to concentrate more on the needs of end-users, with a particular emphasis on the unique requirements and costs associated with serial manufacturing. The development strategy must go beyond the limits of additive manufacturing procedures, including both upstream and downstream approaches. To reduce costs, investments and finance from collaborative partners outside of the process chain must be bolstered.

5.6.2 Putting resources into training, sharing information, and assigning "AM-business visionaries"

Initially, prevailing endeavors to educate engineers in institutions need to continue and amend. In addition, the knowledge of 3D printing needs to be delivered via knowledge transfer projects and commercial training programs. Several studies indicate that companies spent more on skills, knowledge, and people followed by machines. The present hurdle lies in the lacking of professionals inclination to enhance 3D printing processes coupled with earnings in the firm's government financing is an alternative to promote private individual and corporate training in 3D printing. Educational training programs and dedicated transfer initiatives may help to improve industrial knowledge development.

5.6.3 Consolidation and push endeavors in standardization

Standards implementation throughout the industry for instance, in small and medium enterprises, has to be a fundamental objective. Three-dimensional printing standardization committees should continue to take one step further.

Furthermore, the advantages of adopting standards must be communicated to enterprises in a comprehensive and interrelated framework. The government should be enthusiastic to incentivize the research institutions, companies, and industries in the process of standardization to safeguard current state of knowledge via committed funding opportunities.

5.6.4 Boosted and streamlined financing to keep technology readiness research going

The landscape for 3D printing-related processes in technology readiness research has to be improved in partnership with public partners, research institutes, and academic organizations to expedite their industrial applications. 3D printing is often a difficult process that still necessitates much research; nevertheless, if funding opportunities remain unchanged, 3D printing research will dwindle. Streamlining government financing methods is proven to be a hot topic, especially in the areas of cost cutting, quality assurance, and software development.

5.6.5 Implement subsequent sustainability and circular economy

According to many studies, a significant focus on sustainability will benefit 3D printing, which can aid with carbon reduction, for example. Three-dimensional printing may be considered a critical component of a long-term transformation plan for businesses and governments. Incentives such as new rules or financing may steer 3D printing toward a "cradle-to-cradle" and circular economy approach. Furthermore, industry analysts believe that pricing emissions would lead to an increased usage of 3D printing. Companies that combine 3D printing with a comprehensive sustainability plan may create new assets for consumers and stakeholders. 3D printing supports the circular economy by providing novel material alternates, design freedom, sustainable manufacturing, better products, and simplified resource reusability (https://www.additivemanufacturing.media/articles/infographic-how-additive-manufacturing-supports-the-circular-economy).

5.6.6 Bring issues to light and investor quality confirmation and improved capabilities

If nothing is done, qualification might become a deep-seated stumbling block to progress. Extensive research and a push for quality assurance that is often less expensive must be taken into account. Continuous investment in organized collaboration among end-users, norming bodies, regulatory agencies, research, and component suppliers is undertaken to solve this problem. For future industrialization, the qualifying technique must transition from a sweeping a step-by-step method to qualification of a shared process.

5.6.7 Large-scale digitalization of the interaction chain

Consistent data availability and digitization besides the entire process chain needs to be improved as demonstrated by various studies. Hence, appropriate data formats, opening, and allocation of metadata, and standardized interface should be imposed. Machine and software manufacturers are required to tem up and comprehensively enhance the software process chain, especially during consideration of artificial intelligence (AI) capabilities.

A big jump in industrialized 3D printing may be cooperatively developed from inside the 3D printing ecosystem, as seen in the suggested call to action outlined above. With aforesaid changes, enormous new applications and future models will be feasible in the upcoming decade. As a result, the whole market is always moving forward, and various industries have the potential to be disrupted.

5.7 CONCLUSION

Implementation of 3D printing in industries, there is still much to be done. However, this transformative technique has made a quantum leap ahead by creating faster and improved systems and developing novel materials. More attention is focused on broadening the list of approved standards to promote standardization. In addition, continuous attempts are made to bridge the existing knowledge voids and cultivate an advanced generation of AM experts. Lastly, the 3D printing industry is becoming more integrated as the industries are looking forward to collaborating, in a bid to result in extensive solutions. The above-discussed scope of current 3D printing opportunities, fundamental challenges in AM adoption with their solutions, and call of action are indicative of a booming industry that will continue to evolve in the coming years.

REFERENCES

Abdulhameed, O., Al-Ahmari, A., Ameen, W., & Mian, S. H. (2019). Additive manufacturing: Challenges, trends, and applications. *Advances in Mechanical Engineering*, *11*(2), 1687814018822880.

AMFG (2018). Interview: HP's Global Head of Metals on the Impact of HP Metal Jet. Accessed on 28th March 2020.

Basso, M. , Betti, F. , Cronin, I. , Schönfuß, B. , Meboldt, M. , Omidvarkajan, H. , Daniel, L. , Rainer, S. , & Seidel, C. (2022). An additive manufacturing breakthrough: A how-to guide for scaling and overcoming key challenges. White Paper. World Economic Forum.

Campbell, T., Williams, C., Ivanova, O., & Garrett, B. (2011). Could 3D printing change the world Technologies, Potential, and Implications of Additive Manufacturing, Atlantic Council, Washington, DC.

Davies, S. (2020). Joint innovation project develops qualification guidelines for 3D printed heavy industry components. Accessed on 28th March 2020 Deloitte, "The smart factory: responsive, adaptive, connected manufacturing," 2017. Accessed on 28th March 2020.

Deloitte (2017). The smart factory: Responsive, adaptive, connected manufacturing. Accessed on 28th March 2020.

Denkena, B., Dittrich, M. A., Henning, S., & Lindecke, P. (2018). Investigations on a standardized process chain and support structure related rework procedures of SLM manufactured components. *Procedia Manufacturing*, *18*, 50–57.

3DSystems: https://www.3dsystems.com/customer-stories/partial-hand-prosthetics-company-scales-manufacturing-help-3d-systems

Dumitrescu, G. C., & Tanase, I. A. (2016). 3D printing-a new industrial revolution. *Knowledge Horizons. Economics*, *8*(1), 32.

Deutsche Bahn: https://www.bahn.de

Huang, Y., Leu, M. C., Mazumder, J., & Donmez, A. (2015). Additive manufacturing: Current state, future potential, gaps and needs, and recommendations. *Journal of Manufacturing Science and Engineering*, *137*(1), 1–10.

Johnson & Johnson: https://www.jnj.com

Koren, Y. (2006). General RMS characteristics. Comparison with dedicated and flexible systems. Anatoli I. Dashchenko. In *Reconfigurable manufacturing systems and transformable factories* (pp. 27–45). Springer, Berlin, Heidelberg.

Layani, M., Wang, X., & Magdassi, S. (2018). Novel materials for 3D printing by photopolymerization. *Advanced Materials*, *30*(41), 1706344.

Li, P., Mellor, S., Griffin, J., Waelde, C., Hao, L., & Everson, R. (2014). Intellectual property and 3D printing: A case study on 3D chocolate printing. *Journal of Intellectual Property Law & Practice*, *9*(4), 322–332.

Magisetty, R., & Cheekuramelli, N. S. (2019). Additive manufacturing technology empowered complex electromechanical energy conversion devices and transformers. *Applied Materials Today*, *14*, 35–50.

Martinsuo, M., & Luomaranta, T. (2018). Adopting additive manufacturing in SMEs: exploring the challenges and solutions.*Journal of Manufacturing Technology Management*, *29*(6), 937–957.

Mellor, S., Hao, L., & Zhang, D. (2014). Additive manufacturing: A framework for implementation. *International Journal of Production Economics*, *149*, 194–201.

Norrish, J., Polden, J., & Richardson, I. M. (2021). A review of wire arc additive manufacturing: Development, principles, process physics, implementation and current status. *Journal of Physics D: Applied Physics*, *54*(47), 473001.

Oettmeier, K., & Hofmann, E. (2017). Additive manufacturing technology adoption: An empirical analysis of general and supply chain-related determinants. *Journal of Business Economics*, *87*(1), 97–124.

Omidvarkarjan, D., Rosenbauer, R., Klahn, C., & Meboldt, M. (2022). Implementation of additive manufacturing in industry. In *Springer Handbook of Additive Manufacturing*, forthcoming.

Point designs: http://www.pointdesignsllc.com/

Punia, U., Kaushik, A., Garg, R. K., Chhabra, D., & Sharma, A. (2022). 3D printable biomaterials for dental restoration: A systematic review. *Materials Today: Proceedings*, *63*, 566–572.

Quinlan, H., & Hart, A. J. (2020). Additive manufacturing: Implications for technological change, workforce development, and the product lifecycle. Research Brief RB14-2020, November 24 (2020).

Schneider, D. (2006). FAB: The Coming Revolution on Your Desktop–From Personal Computers to Personal Fabrication. *American Scientist*, *94*(3), 284–285.

Sharma, A., Chhabra, D., Sahdev, R., Kaushik, A., & Punia, U. (2022). Investigation of wear rate of FDM printed TPU, ASA and multi-material parts using heuristic GANN tool. *Materials Today: Proceedings*, *63*, 559–565.

Sonar, H., Khanzode, V., & Akarte, M. (2020). Investigating additive manufacturing implementation factors using integrated ISM-MICMAC approach. *Rapid Prototyping Journal*, *26*(10), 1837–1851.

TÜV SÜD: https://www.tuvsud.com/en-gb/about-us

Ukobitz, D. V. (2020). Organizational adoption of 3D printing technology: A semi-systematic literature review. *Journal of Manufacturing Technology Management*, *32*(9), 48–74.

Wohlers, T. (2013). Wohler's report 2013.

https://www.additivemanufacturing.media/articles/infographic-how-additive-manufacturing-supports-the-circular-economy

Chapter 6

Fabrication of Light Metal Alloys by Additive Manufacturing

Binnur Sagbas and Özgür Poyraz

CONTENTS

6.1 INTRODUCTION

In the last decades, the world has been facing environmental challenges as the result of population growth and industrial development. Various studies were conducted on the emerging environmental issues, in order to comprehend the details, to identify possible measures or solutions, and to raise the awareness of the individuals. Among different sectors, there are also studies on the manufacturing industry, with a focus on developing more sustainable processes in line with awareness of emerging environmental issues. Sustainable manufacturing term corresponds to decrease waste of materials, chips and scraps (Sreenivasan et al., 2010). Moreover, decreasing energy consumption and environmental pollution by reducing the cutting fluids, oils and related substances, which are widely used for conventional manufacturing processes might contribute to sustainable manufacturing.

Additive manufacturing (AM) technologies, which enable the production of three-dimensional (3D) parts in a layerwise manner, provide the opportunity to produce parts by minimum waste of materials and energy consumption without the need for any die, tool, fixture, cutting fluids or oil (Yang et al., 2020a; Ford and Despeisse, 2016; Peng et al., 2018). Furthermore, the ability of producing 3D geometries with complex shapes allow implementation of topology optimization, generative design and lattice structures. As a result, a functional part can be manufactured with less material and its weight can be reduced without sacrificing strength (Prathyusha and Babu, 2022). The reduction of weight and thus the energy consumption is of great importance especially for the aerospace, maritime and automotive industries. The adoption of lighter parts and components provides a considerable amount of fuel efficiency which results in decreasing the carbon emissions and the related environmental impacts. For this reason, lightweight metallic materials such as aluminum and titanium alloys are of great interest for the aforementioned industries. Besides, their high specific strength, good corrosion resistance and easy processability in powder form make them widely preferred materials for additive manufacturing applications particularly in powder bed fusion (PBF) processes.

Nowadays, with the rapid development of the AM technologies, it is possible to produce near-net-shape (NNS) functional parts by laser powder bed fusion (L-PBF) processes from light metals such as AlSi10Mg and Ti6Al4V powders. However, because of the complex physical phenomena during rapid solidification of the L-PBF processes, properties of the final product highly depend on the process parameters which have to be optimized according to the material, design and geometry of the part (Ghio and Cerri, 2022). Therefore, L-PBF process parameters of light weight metals, such as laser power, beam diameter, scanning speed, layer thickness, hatch distance and their effects on the final product properties, are explained and discussed in this chapter. The rest of the chapter is organized as follows. First section explains the principles of L-PBF process followed by the detailed explanations of the essential process parameters. Afterward, each section deals with the parameter values used for the relevant light metal alloy and the influence of these values on the physical, mechanical, surface and dimensional attributes of the parts. In the last section, the entire chapter is summarized and the most influential parameters for L-PBF of light metal alloys are highlighted.

6.2 LASER POWDER BED FUSION ADDITIVE MANUFACTURING

The laser powder bed fusion (L-PBF) processes depend on the principle of melting and fusing metal powders with the utilization of the laser power as the heat source. Three-dimensional part geometry is produced layer-upon-layer on the build platform. As the name suggests, the powder bed is formed by recoating the dispensed powder at every layer to the build chamber until the full part is produced. As the height of the full part is achieved by elevating the build platform along Z-axis, the cross-sections at each layer are formed by moving the laser beam along the X-axis and Y-axis with the help of galvanometric mirrors. While the potential risks of introducing undesired elements to the composition of the produced parts are eliminated via processing under inert gas atmosphere, other risks such as as-received humidity

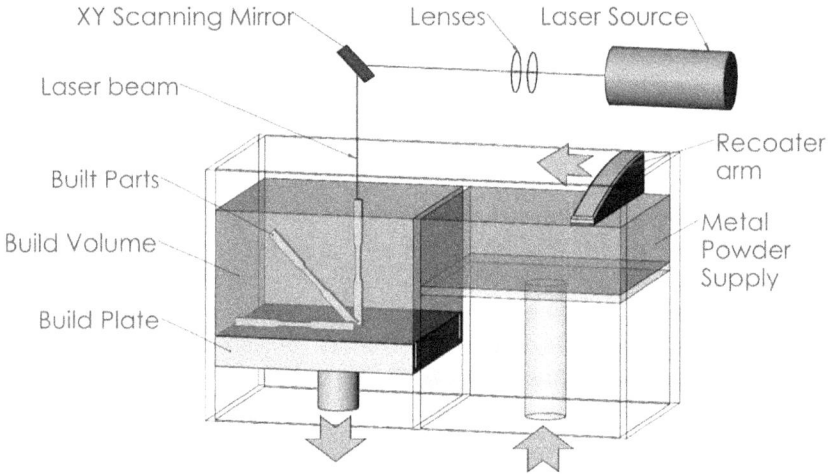

Figure 6.1 Schematic drawing of L-PBF system (Under Creative Commons CC BY 4.0 license).
Source: Mooney and Kourousis (2020).

of powders are avoid by heating the build platform. A schematic representation of the generic L-PBF system can be seen in Figure 6.1.

Even though it is beneficial for researchers and engineers to grasp the details of L-PBF machines and their working principles, further knowledge is needed in order to produce high-quality parts together with secure microstructure and mechanical properties (Yasa and Poyraz, 2020). This knowledge then can be used to optimize various process parameters which lead to complex multi-physics phenomena during rapid solidification (Ahmed et al., 2022).

Powder bed fusion processes have various parameters, which can be classified in four basic groups such as laser-based parameters, powder-based parameters, scan-based parameters and temperature-based parameters (Figure 6.2). Laser based parameters consist of laser power, laser beam

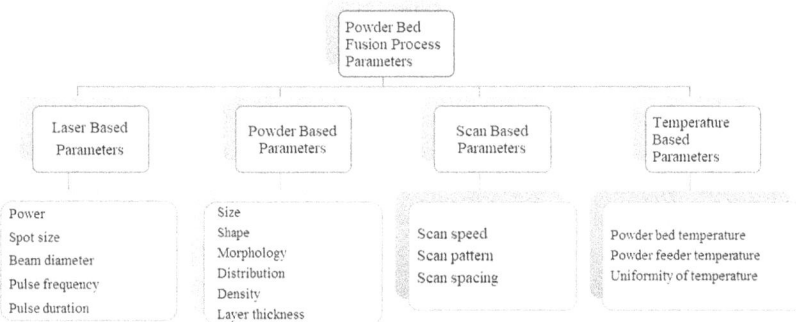

Figure 6.2 Classification of powder bed fusion process parameters.

diameter, frequency of the laser pulse and its duration. Particle size, distribution, sphericity, density and morphology can be classified under powder-based parameters (Gibson et al., 2021).

Laser power is the most important laser-based parameters, which highly dominate the heat input to molten the powder, fusion process and molten pool characteristics. Higher laser power provides higher heat energy, wider molten pool with enough depth which guarantees to melt and fuse the metal powder with the previous layer. This phenomenon results in compact structured, highly dense parts without pores and gaps (Sagbas et al., 2021). It is important to note that achieving optimum molten pool width and depth is the key factor for obtaining highly dense parts without pores and gaps. The criterion presented with Equation 6.1 is offered by Tang et al. in order to achieve full melting (Tang et al., 2017).

$$\left(\frac{h}{W}\right)^2 + \left(\frac{t}{D}\right)^2 \leq 1 \tag{6.1}$$

where W is the melt pool width (mm), h is the hatch distance (scan spacing) (mm) refers to the distance between the center of two following laser scan track, t is the layer thickness (mm) and D is the depth of the molten pool (mm).

Melt pools with dimensions greater than optimum values will cause unstable behavior of the molten material by capillary (Rayleigh) and Marangoni effect. Viscosity, capillarity and surface tension gradient determine the flow of the molten metal. Under the unstable conditions, it would be difficult to control the direction of the melt flow, which causes unexpected microstructures, mechanical properties and surface quality. Moreover, higher depth of the melt pool will cause high heat input to the previously, already solidified layer and change its microstructures by uncontrollable thermal gradients. As a result, mechanical properties, surface quality and dimensional accuracy will be affected by microstructure changes and residual stresses (Sagbas et al., 2021). Figure 6.3 shows the schematic representation of the melt pool and the corresponding geometric characteristics.

It is important to note that laser power is not a unique parameter for controlling the melt pool dimensions and its stability. There are many different parameters, beside whose individual effects, most of them are strongly interrelated with each other and have combined effects on product properties. For expressing this combined effect, input energy density (E; J/mm^3) term, which can be expressed by the most effective L-PBF parameters such as laser power (P; W), scan speed (v; mm/s), hatch distance (h; mm) and layer thickness (t; mm), is defined with the Equation 6.2. Energy density represents the energy delivered to a unit volume of powder material (Brandt, 2016).

$$E = \frac{P}{v.\,h.\,t} \tag{6.2}$$

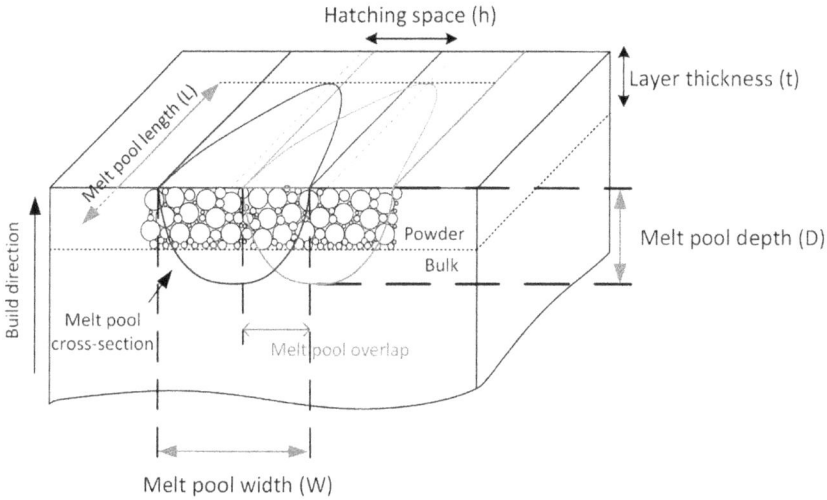

Figure 6.3 Schematic representation of the melt pool and the corresponding geometric characteristics (Under Creative Commons CC BY 4.0 license).

Source: Letenneur et al. (2019).

This complex interaction between the L-PBF process parameters may result in different product properties in terms of dimensional accuracy, surface quality, mechanical properties, density and porosity. By optimization of these parameters, it is possible to obtain desired product specification otherwise unexpected parameter interactions may cause different kinds of defects such as incomplete melting, pinholes, residual stresses, undesired pores, form deviations, rough surfaces and low mechanical properties. For instance, lower values of laser energy density with low laser power and high scanning speed result in insufficient melting of metal powder, lack of fusion and so high porosity. On the contrary, increasing the laser power with a constant scanning speed will cause excessive heating and deeper laser penetration, which result formation of keyholes and re-melting or microstructural changes of the previous solid layers (Kamath et al., 2014). While both laser power and scanning speed increased to a certain limit, it will result in the formation of unstable molten pool behavior with vaporization, micro-segregation of the alloying elements, balling effect and discontinuous melt tacks(Ahmed et al., 2022).

Hatch distance highly affects the stability of the melt pool and micro-structure of the manufactured part. While low hatch distance is used, large overlaps will occur and cause high energy input. Stable melt pool will occur at the optimal hatch distance (Figure 6.4). If the hatch distance is less than the optimal value, an unstable melt pool will occur because of the excessive heat input. On the contrary, while the hatch distance is large, it will cause inadequate heat input and melting so there will be gaps between the scans.

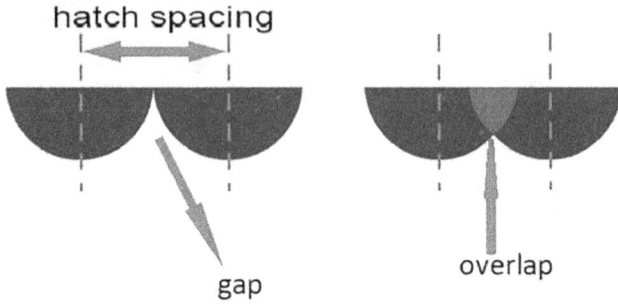

Figure 6.4 Representative drawing of hatch spacing and overlap.

Source: Sagbas et al. (2021).

According to the expected product specifications, the hatch distance parameter must be optimized (Sagbas et al., 2021).

Beside scan speed, scanning strategies also play an important role in heat dissipation required for bonding of powder. Scanning strategies designate the heat transfer, solidification processes and thermal gradients between layers which affect microstructure, mechanical properties, residual stresses and so dimensional accuracy and surface quality of the L-PBF manufactured part (Yasa and Poyraz, 2021). Various type of scanning strategies used for L-PBF are shown in Figure 6.5. Appropriate hatch distance, laser scanning patterns and layer thickness ensure enough heat dissipation required for bonding between powders and layers.

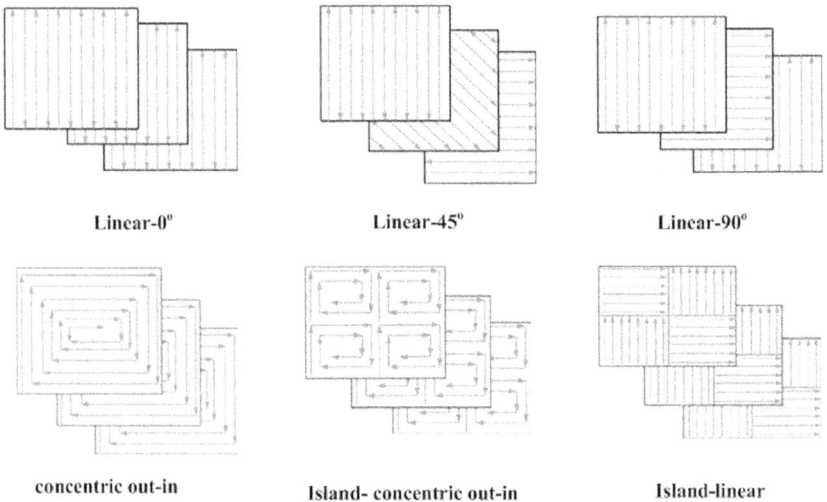

Figure 6.5 Various type of scanning strategies used for L-PBF (Under Creative Commons CC BY 4.0 license).

Source: Abdelmoula et al. (2022).

Powder-based parameters have also an important effect on final properties of the L-PBF manufactured parts (Srivatsan and Sudarshan, 2015). Size and shape of the powder highly affect the density and behavior of the powder during the fusing process. Spherical powders with high surface smoothness provide higher flowability while finer powders mixed with larger ones may fill the gaps between powders and increase packing density (Zhu et al., 2007; Rajabi et al., 2008). Thermal conductivity and solidification rate of the powder increases proportionally by increasing packing density. As a result, finer microstructures which provide higher mechanical properties can be achieved.

Parameter levels and their effects may differ according to the material type because of their chemical and physical properties such as compositions, thermal properties, surface energy, density and laser absorption. In the following sections, L-PBF parameters and their effects are discussed on most widely used light metal alloys of aluminum and titanium with the reference based on previous studies.

6.3 LASER POWDER BED FUSION ADDITIVE MANUFACTURING OF TITANIUM ALLOYS

Although more than two centuries have passed since the elemental discoveries of titanium, its industrial use could only begin in the second half of the 20th century due to the difficulties in purification processes (Leyens and Peters, 2003). The production process of titanium, which consists of four main stages in general, includes the reduction of titanium ore to obtain a sponge-like structure followed my melting to form an ingot (Lütjering and Williams, 2007). Later on, general mill products such as billet, bar or plate are produced by converting the ingot via primary fabrication. Finally finished products are obtained through secondary fabrication techniques. Among these stages, different solutions were sought especially for the initial reduction. With the utilization of early developed Kroll process, titanium tetrachloride was used instead of titanium dioxide during reduction, and this has a complex process route in which magnesium is also involved (Leyens and Peters, 2003). Titanium alloys, which were made feasible for industrial use with other fabrication techniques and additions developed later, have expanded their application areas in less than half a century and have become preferred for several industries like aviation, biomedical, dentistry, automotive and machinery.

Despite the initial fabrication difficulties, titanium alloys undoubtedly owed their widespread use to the different advantages they offer. Among these advantages, their mechanical performance comes to the fore at first glance, and in this sense, the specific strength, or in other words, the high strength-to-mass ratio is one of the most desirable features for applications where weight reduction is important, especially in the aerospace industry

(Bray, 1990). In addition to the advantageous properties they offer, the wide variety of alloy types and manufacturing methods has an important role behind the widespread use of titanium. In this context, titanium alloys which have many available grades such as commercially pure, alpha, beta or alpha-beta can serve different purposes. A similar diversity exists in terms of production methods. Accordingly and depending on the grade, semi-finished titanium alloys can be processed into their final form by conventional methods such as forging, casting, welding and machining, as well as by unconventional methods such as electro-chemical processing, superplastic forming and additive manufacturing (Benedict, 2017; Bourell et al., 2020).

Thanks to the design freedom offered by additive manufacturing technologies, allowing internal channels, lattice structures and topologically optimized geometries, application of titanium alloys, which already have a great advantage in terms of strength-to-mass ratio, is very compelling for industries like aerospace. Powder bed fusion (PBF) additive manufacturing, which is the most mature group in terms of the production of metallic materials among all additive manufacturing families defined by international standards (AM-POWER, 2020), includes a wide variety of applications available to titanium alloys. These applications provide significant benefits in terms of both design and production if performed correctly. However, incorrect applications can lead to undesirable results.

Among the relevant PBF groups, the most common is the laser powder bed fusion (L-PBF) additive manufacturing. Today, the popular titanium grades for additive manufacturing research with L-PBF are cp-Ti (Grade 1), cp-Ti (Grade 2), cp-Ti (Grade 3), Ti6Al4V (Grade 5) and Ti6Al4V - ELI (Grade 23). Certain international standards for these grades have been published and the F3302-18 standard provides general information about Grade 3, Grade 5 and Grade 23, while the F2924-14 standard also includes additional information about Grade 5 and F3001-14 standard Grade 23 (ASTM International, 2018, 2014a, 2014a). The first thing to be considered in terms of L-PBF of titanium alloys is the raw material used. The chemical composition of the powders for the production of titanium alloys via L-PBF is presented at Table 6.1 based on the international standards.

Another point to be considered in terms of powder materials for L-PBF additive manufacturing of titanium alloys is powder particle size, distribution and powder morphology. While the powder particle sizes with average diameter of 40 μm expected to have a gaussian distribution that can go up to 100 μm, morphological considerations include sphericity and satellite free pieces. ASTM International (ASTM International, 2014b) refers to two common methods which can be employed to identify the chemical composition of the alloying elements of metal powder other than oxygen, nitrogen, hydrogen and carbon. These methods are Atomic Emission Plasma Spectrometry and Optical Emission Spectroscopy. Powder morphology, which is another important characteristic, is widely evaluated by qualitative

Table 6.1 Chemical composition of the common titanium and alloy powders (ASTM International, 2018)

Alloy	Carbon, Max	Oxygen, Max	Nitrogen, Max	Hydrogen, Max	Iron, Max	Aluminum	Vanadium	Yttrium	Other, Max
cp-Ti (Grade 3)	0.08	0.35	0.05	0.015	0.30	-	-	-	0.5
Ti6Al4V (Grade 5)	0.08	0.20	0.05	0.015	0.30	5.50–6.75	3.50–4.50	0.005	0.5
Ti6Al4V -ELI (Grade 23)	0.08	0.13	0.05	0.015	0.25	5.50–6.50	3.50–4.50	0.005	0.5

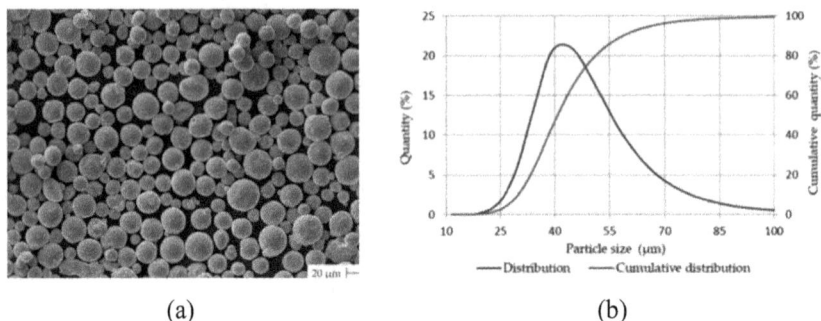

(a)   (b)

Figure 6.6 a) Morphological evaluation, and b) particle size analysis of Ti6Al4V powder for
L-PBF (Under Creative Commons CC BY 4.0 license).

Source: Malý et al. (2019).

definitions and/or particle shapes. Image analysis or light scattering methods (ASTM International, 2014b) can be employed for the relevant characterization tasks. These qualitative definitions include the sphericity of the powders and aspect ratio of the outer dimensions. Particle size distribution (PSD) of metal powders can be characterized using several methods as sieving, image analysis and light scattering (Laser light diffraction) (Figure 6.6) (ASTM International, 2014b).

After providing a suitable material in terms of standards and verifying this, energy density parameters are one of the first subjects to focus on in the process development of titanium alloys. As in other alloys, the energy density parameters are obtained by dividing the laser power by the layer thickness, scanning speed and hatch distance, and the effect of the parameter combinations on the part density or porosity is examined. For the porosity investigations, in addition to density measurement with Archimedes method, image processing over microstructure or non-destructive characterizations with X-CT are employed (Slotwinski et al., 2014; Arvieu et al., 2020). Although the Archimedes method is preferred more because of its low investment cost and non-destructive nature, there is a risk of lower density determination compared to other methods (Mahmud et al., 2021). Among the reasons for this situation may be the fact that there is a certain level of roughness on the surfaces of additively manufactured titanium alloys and that the liquid used in the Archimedes method cannot enter between them, so the measured volume is higher, and the density is lower. While lack of fusion errors are encountered in titanium alloys where the energy input is low, energy input increase lead to other porosities due to inert gas entrapment, inert gas dissolved in liquid melt and released on solidification, and metal gas evaporated by high laser energies and re-solidification (Vilaro et al., 2011; Gong et al., 2014). In a study by Kasperovich et al., it was determined that among different energy input parameters, scanning speed has a major effect on porosity in both directions and laser power in the direction of lack of fusion (Kasperovich et al., 2016). In the aforementioned study, the hatch distance was considered as the least effective parameter since the laser melting area was large (Kasperovich et al., 2016). Another study confirms that laser power and laser scanning speed were the parameters with the highest effect on porosity of titanium alloys (Malý et al., 2019). In a study in which various publications are reviewed, it is seen that the energy density range that minimizes the porosity for Ti6Al4V is at the level of 85-105 J/mm^3 (Poyraz et al., 2021). In addition, in another study compiling the literature on cp-Ti parts, it was revealed that the appropriate energy density range for this material is 95-120 J/mm^3 (Zhang et al., 2016).

In terms of the parts whose energy density is optimized and can reach the maximum possible part density, the expected mechanical strength values according to international standards are 380MPa yield strength (YS) and 450MPa ultimate tensile strength (UTS) for cp-Ti, 825MPa YS and 895MPa UTS for Ti6Al4V, 795MPa YS and 860MPa UTS for Ti6Al4V ELI (ASTM International, 2018). These values are expected from the parts produced in the X-Y and Z directions. However, it has been seen in the literature that 555MPa YS and 757MPa UTS are achievable for cp-Ti (Attar et al., 2014). Again, in the literature, it has been determined that YS and UTS values exceed 1100 MPa in studies with Ti6al4V-ELI (Mahmud et al., 2021).

When titanium alloys are examined in terms of mechanical properties, it has been seen that secondary processes have an important effect on these properties, in addition to factors such as the process parameters used. In this respect, different publications assess the room temperature mechanical strength of as built (AB), stress relieved (SR), heat treated (HT) and hot isostatic pressed (HIP) specimens. As can be seen from Figure 6.2, while the highest strength was in the AB state, the lowest strength was in HIP for Ti6Al4V following to L-PBF. However, the effects of HIP should not be evaluated solely by considering at the YS and UTS values. When the percentage of elongation values are considered, it is observed that AB samples can decrease down to 6%, whereas HIP samples can go up to 19% levels (Rekedal and Liu, 2015; Cain et al., 2015; Greitemeier et al., 2016; Kasperovich and Hausmann, 2015; Leuders et al., 2013). It is seen that another advantage of the HIP process is in the value of fracture toughness. While the fracture toughness value of AB made samples is 28 MPa√m in the highest measured directions, this value goes up to 57.8 MPa√m in HIP made samples (Rekedal and Liu, 2015; Cain et al., 2015; Greitemeier et al., 2016; Kasperovich and Hausmann, 2015; Leuders et al., 2013). Figure 6.7 shows Yield and ultimate tensile strength of Ti6Al4V under different secondary process conditions following to L-PBF.

Additional to mechanical attributes of titanium alloys fabricated by L-PBF, it is also important to consider the influence of processing parameters on surface quality (Oyesola et al., 2021). Studies of various researchers show that laser power and scanning speed affects the quality with respect to roughness and hardness of the produced part (Bourell et al., 2020). As the

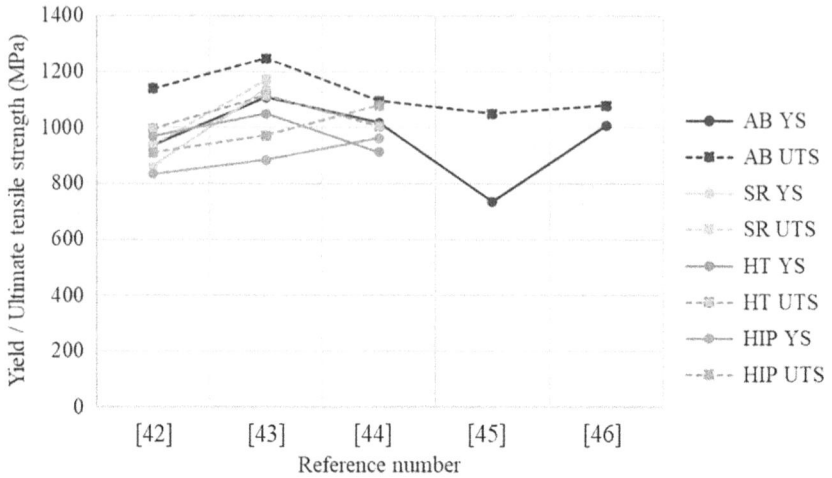

Figure 6.7 Yield and ultimate tensile strength of Ti6Al4V under different secondary process conditions following to L-PBF.

optimum process parameters were laser power was determined as 300 W and scan speed as 1400 mm/s. This set of process parameters produce a minimum surface roughness of 13.006 μm for the top surface and 62.166 μm for the side surface (Bourell et al., 2020).

6.4 LASER POWDER BED FUSION ADDITIVE MANUFACTURING OF ALUMINUM ALLOYS

By having high melt fluidity and ease in manufacturing, AlSi10Mg alloy is the most widely used alloy for L-PBF processes. Due to its high strength, thermal conductivity, low thermal expansion, high load-bearing properties and high reflectivity, it is mainly used in automotive and aerospace industries (Pola et al., 2019). AlSi10Mg comprises various elements, including titanium, aluminum, silicon, copper, iron, magnesium, manganese, zinc and some residuals. Chemical composition of AlSi10Mg is given in Table 6.2 and scanning electron microscopy (SEM) image in Figure 6.8 (Ishfaq et al., 2021; Sagbas, 2020).

L-PBF process parameters have great effect on the above mentioned AlSi10Mg part properties. Mamoon et al. studied the effect of process parameters on surface quality of LPBF manufactured AlSi10Mg and Al6061

Table 6.2 Chemical composition of AlSi10Mg (Source: Ishfaq et al., 2021)

Al (%)	Si (%)	Fe (%)	Cu (%)	Mn (%)	Mg (%)	Ni (%)	Zn (%)	Pb (%)	Sn (%)	Ti (%)
Balance	9.0–11.0	0.55	0.05	0.45	0.2–0.45	0.05	0.10	0.05	0.05	0.15

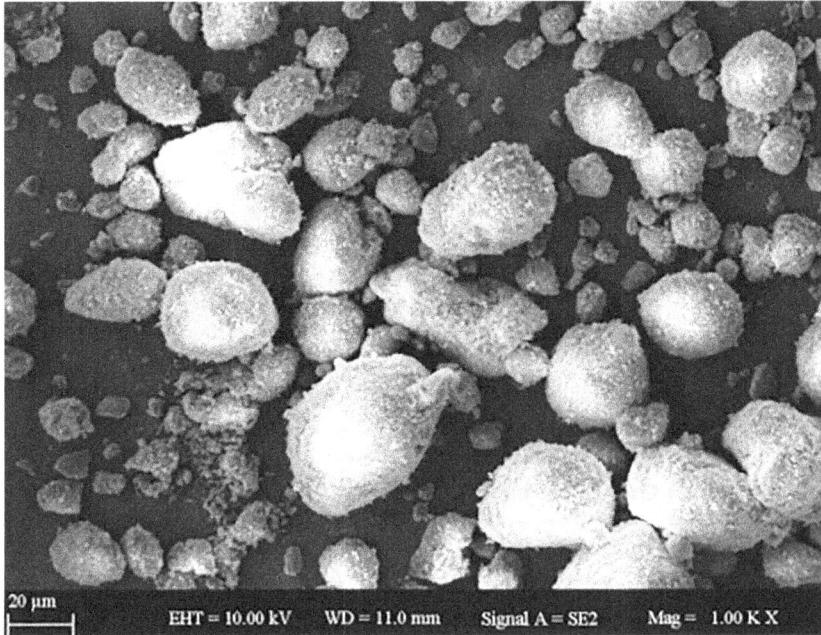

Figure 6.8 SEM image of AlSi10Mg.

Source: Sagbas (2020).

parts (Maamoun et al., 2018). They tested eight samples manufactured with different levels of laser power, scan speed and hatch distance which provide different energy densities, ranges between 27-65 J/mm^3. They reported that the best surface quality with 4.5 μm roughness value was achieved with the highest energy density 65 J/mm^3 (Maamoun et al., 2018). Dimensional accuracy of the L-PBF manufactured parts are also highly affected by process parameters. Mukherje et al. reported that thermal distortions, aroused by large melt pool volumes, caused large dimensional deviations (Mukherjee et al., 2018). Bagci et al. studied the dimensional accuracy of L-PBF of AlSi10Mg functional parts. They concluded that besides laser power, scan speed and scanning angle, beam offset is an important parameter that affects dimensional accuracy of screw teeth as it is shown in Figure 6.9 (Bagci et al., 2020).

Mechanical properties of L-PBF manufactured parts are mostly the result of process parameters. Complex interactions between powder, heat source and parameters itself cause complex thermal gradients which result in different mechanical properties by gas pores, keyholes, incomplete melting, residual stresses and different microstructures. Gasses, that trapped inside the powder particles while chemical modifications are applied to the powder, may release at melting step and remain during solidification. As a result, gas pores which have important effect on mechanical properties, may be aroused (Ferro et al., 2020; Sames et al., 2016).

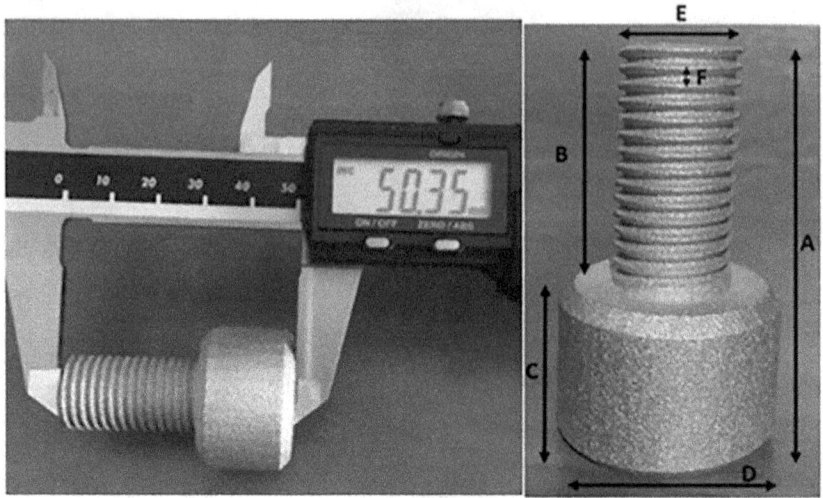

Figure 6.9 Dimensional accuracy measurement of L-PBF manufactured AlSi10Mg functional parts.

Yang et al. investigated laser power, linear energy density (J/mm) and volumetric energy density (J/mm³) on pore formation, microstructural and mechanical properties of L-PBF fabricated AlSi10Mg parts (Yang et al., 2020b). They reported that with the 111.1 J/mm³ constant volumetric energy densities, the pore quantities were reduced by decreasing linear energy density from 0.2 J/mm to 0.15 J/mm, and the ultimate tensile strength increased to 475.18 MPa. The tensile strength decreased to 422.55 MPa while the linear energy density decreased to 0.12 J/mm. The authors concluded this situation as a result of formation of some large irregular pores (larger than 100μm) because of insufficient linear energy density. Moreover, while the linear and volumetric energy densities were kept constant at 0.2 J/mm and 111.1 J/mm³ levels, the ultimate tensile strength increased by decreasing the number of pores with increasing laser power and scanning speed (Yang et al., 2020b).

However, pores are generally formed by gas, wrapped into the gas atomized powder or entrapped in build chamber during L-PBF process, for AlSi10Mg parts, the gas pores may also be a result of reduction of H_2O with the following reactions which occur due to the thermal cycles during the process. The hydrogen diffuses into the molten aluminum in monoatomic form and may cause formation of pores (Ghio and Cerri, 2022).

$$2H_2O \rightarrow 2H_2 + O_2$$

$$3H_2O + 2Al \rightarrow Al_2O_3 + 3H_2$$

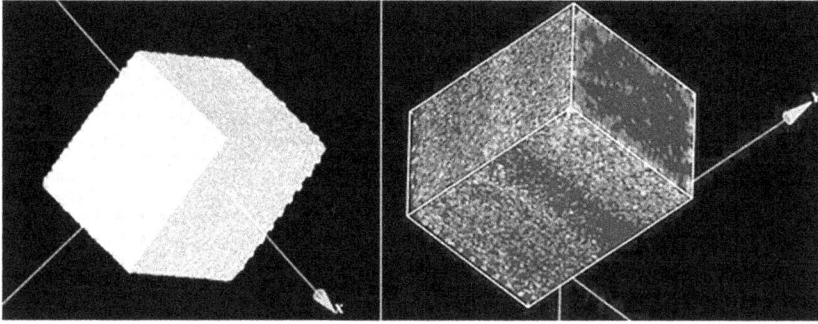

Figure 6.10 Measurement of pores in LPBF manufactured AlSi10Mg part by micro computed tomography.

Source: Sagbas and Durakbasa (2019).

Metrological assessment of the size, shape, numbers and dispersion of the pores are important for predicting mechanical properties of the parts. Archimedes method is generally used for defining density and porosity of the parts; however, this method doesn't give information about size and dispersion of the pores. Micro computed tomography provides opportunity to evaluate volumetric fraction and dispersion of the pores while microscopic techniques best choice for size evaluation (Figure 6.10) (Sagbas and Durakbasa, 2019).

Another important parameter that affects product properties is build platform temperature. By affecting solidification, cooling rate and thermal gradient during manufacturing of a part, it causes serious changes in the microstructure, residual stresses and so mechanical properties. Macias et al. studied about the build platform temperature, and they set the temperature 30 °C to 200 °C (Macías et al., 2020). They concluded that 200 °C build platform temperature decreased the residual stress approximately about negligible level but the strength was lower with coarser grains which were prone to damage (Macías et al., 2020). Influence on microstructure, strength and ductility of build platform temperature during laser powder bed fusion of AlSi10Mg were also investigated by other researchers. In a study on build platform temperature conducted by Awd et al., authors focused on the fatigue behavior of AlSi10Mg parts for which the cooling rates were controlled by heating the build platform up to 200°C (Awd et al., 2019). They concluded that controlling the cooling process provided homogeneous microstructure and modified pore structures and forms. As a result, higher fatigue strength was recorded for the parts manufactured on the heated build platform (Awd et al., 2019).

Scan strategy is another important parameter that affects density, surface quality and microhardness of the L-PBF manufactured parts. Bhardwaj et al. reported some effects of scanning strategies on the density (Bhardwaj and Shukla, 2018). Moreover, they concluded that surface roughness of one-way scanned samples was about 2 times higher than XY scanned ones because there were more voids and cracks in one-way scanning than XY scanning. Therefore, XY scanning strategy is recommended for high quality finished parts (Bhardwaj and Shukla, 2018).

Han et al. focused on the effect of hatch distance and scanning speed on dimensional accuracy and surface quality of L-PBF fabricated parts (Han et al., 2018). They reported that there was an inflection point for scanning speed while hatch distance was constant and dimensional accuracy of the part depends on the inflection point. When the scanning speed was under this inflection point, dimensional accuracy increased significantly by increasing scanning speed (Han et al., 2018).

Beside process parameters, post processing has also an important effect on the product quality. Thermal, mechanical, chemical and laser-based post processes are generally applied to the AlSi10Mg L-PBF fabricated parts according to the desired quality specifications. Sagbas applied three different mechanical post processes such as shot peening, blasting and polishing on to the AlSi10Mg parts. The author reported that shot peening caused plastic deformation of the surface and increased micro hardness and wear resistance, while polishing provided the lowest surface roughness (Sagbas, 2020). In another study, Pola et al. (2019) focused on the effect of shot peening on fatigue behavior of AlSi10Mg parts manufactured by L-PBF. They concluded that shot peening considerably increased fatigue strength of the parts.

REFERENCES

Abdelmoula, M., Küçüktürk, G., Juste, E., & Petit, F. (2022). Powder bed selective laser processing of alumina: Scanning strategies investigation. Applied Sciences, 12(2), 764.

Ahmed, N., Barsoum, I., Haidemenopoulos, G., & Al-Rub, R. A. (2022). Process parameter selection and optimization of laser powder bed fusion for 316L stainless steel: A review. Journal of Manufacturing Processes, 75, 415–434.

AM-POWER (2020) Additive Manufacturing New Metal Technologies.

Arvieu, C., Galy, C., Le Guen, E., & Lacoste, E. (2020). Relative density of SLM-produced aluminum alloy parts: Interpretation of results. Journal of Manufacturing and Materials Processing, 4(3), 83.

ASTM International (2018) Additive Manufacturing – Finished Part Properties –Standard Specification for Titanium Alloys via Powder BedFusion, DOI: 10.1520/F3302-18

ASTM International (2014a) Additive Manufacturing Titanium-6 Aluminum-4 Vanadiumwith Powder Bed Fusion, DOI: 10.1520/F2924-14

ASTM International (2014a) Additive Manufacturing Titanium-6 Aluminum-4 VanadiumELI (Extra Low Interstitial) with Powder Bed Fusion, DOI: 10.1520/F3001-14

ASTM International (2014b) Standard Guide for Characterizing Properties of Metal Powders Used for Additive Manufacturing Processes.

Attar, H., Calin, M., Zhang, L. C., Scudino, S., & Eckert, J. (2014). Manufacture by selective laser melting and mechanical behavior of commercially pure titanium. Materials Science and Engineering: A, 593, 170–177.

Awd, M., Siddique, S., Johannsen, J., Emmelmann, C., & Walther, F. (2019). Very high-cycle fatigue properties and microstructural damage mechanisms of selective laser melted AlSi10Mg alloy. International Journal of Fatigue, 124, 55–69.

Bagci, H. B., Sagbas, B., & Durakbasa, M. N. (2020). Effect Of Direct Metal Laser Sintering Process Parameters On Dimensional Accuracy Of AlSi10Mg Parts. DAAAM International Scientific Book, 171–180.

Benedict, G. F. (2017). Nontraditional manufacturing processes. CRC press.

Bhardwaj, T., & Shukla, M. (2018). Effect of laser scanning strategies on texture, physical and mechanical properties of laser sintered maraging steel. Materials Science and Engineering: A, 734, 102–109.

Bourell, D. L., Frazier, W. E., Kuhn, H. A., & Seifi, M. (Eds.). (2020). ASM handbook: Additive manufacturing processes. ASM International.

Brandt, M. (Ed.). (2016). Laser additive manufacturing: Materials, design, technologies, and applications.

Bray, J. W. (1990). Properties and selection: Nonferrous alloys and special purpose materials. ASM Metals Handbook, 2, 1886–1915.

Cain, V., Thijs, L., Van Humbeeck, J., Van Hooreweder, B., & Knutsen, R. (2015). Crack propagation and fracture toughness of Ti6Al4V alloy produced by selective laser melting. Additive Manufacturing, 5, 68–76.

Ferro, P., Meneghello, R., Savio, G., & Berto, F. (2020). A modified volumetric energy density–based approach for porosity assessment in additive manufacturing process design. The International Journal of Advanced Manufacturing Technology, 110(7), 1911–1921.

Ford, S., & Despeisse, M. (2016). Additive manufacturing and sustainability: An exploratory study of the advantages and challenges. Journal of cleaner Production, 137, 1573–1587.

Ghio, E., & Cerri, E. (2022). Additive manufacturing of AlSi10Mg and Ti6Al4V lightweight alloys via laser powder bed fusion: A review of heat treatments effects. Materials, 15(6), 2047.

Gibson, I., Rosen, D. W., Stucker, B., Khorasani, M., Rosen, D., Stucker, B., & Khorasani, M. (2021). Additive manufacturing technologies (Vol. 17). Cham, Switzerland: Springer.

Gong, H., Rafi, K., Gu, H., Starr, T., & Stucker, B. (2014). Analysis of defect generation in Ti–6Al–4V parts made using powder bed fusion additive manufacturing processes. Additive Manufacturing, 1, 87–98.

Greitemeier, D., Dalle Donne, C., Syassen, F., Eufinger, J., & Melz, T. (2016). Effect of surface roughness on fatigue performance of additive manufactured Ti–6Al–4V. Materials Science and Technology, 32(7), 629–634.

Han, X., Zhu, H., Nie, X., Wang, G., & Zeng, X. (2018). Investigation on selective laser melting AlSi10Mg cellular lattice strut: Molten pool morphology, surface roughness and dimensional accuracy. Materials, 11(3), 392.

Ishfaq, K., Abdullah, M., & Mahmood, M. A. (2021). A state-of-the-art direct metal laser sintering of Ti6Al4V and AlSi10Mg alloys: Surface roughness, tensile strength, fatigue strength and microstructure. Optics & Laser Technology, 143, 107366.

Kamath, C., El-Dasher, B., Gallegos, G. F., King, W. E., & Sisto, A. (2014). Density of additively-manufactured, 316L SS parts using laser powder-bed fusion at powers up to 400 W. The International Journal of Advanced Manufacturing Technology, 74(1), 65–78.

Kasperovich, G., & Hausmann, J. (2015). Improvement of fatigue resistance and ductility of TiAl6V4 processed by selective laser melting. Journal of Materials Processing Technology, 220, 202–214.

Kasperovich, G., Haubrich, J., Gussone, J., & Requena, G. (2016). Correlation between porosity and processing parameters in TiAl6V4 produced by selective laser melting. Materials & Design, 105, 160–170.

Letenneur, M., Kreitcberg, A., & Brailovski, V. (2019). Optimization of laser powder bed fusion processing using a combination of melt pool modeling and design of experiment approaches: density control. Journal of Manufacturing and Materials Processing, 3(1), 21.

Leuders, S., Thöne, M., Riemer, A., Niendorf, T., Tröster, T., Richard, H. A., & Maier, H. J. (2013). On the mechanical behaviour of titanium alloy TiAl6V4 manufactured by selective laser melting: Fatigue resistance and crack growth performance. International Journal of Fatigue, 48, 300–307.

Leyens, C., & Peters, M. (Eds.). (2003). Titanium and titanium alloys: Fundamentals and applications. John Wiley & Sons.

Lütjering, G., & Williams, J. C. (2007). Fundamental aspects. Titanium, 15–52.

Maamoun, A. H., Xue, Y. F., Elbestawi, M. A., & Veldhuis, S. C. (2018). Effect of selective laser melting process parameters on the quality of al alloy parts: Powder characterization, density, surface roughness, and dimensional accuracy. Materials, 11(12), 2343.

Macías, J. G. S., Douillard, T., Zhao, L., Maire, E., Pyka, G., & Simar, A. (2020). Influence on microstructure, strength and ductility of build platform temperature during laser powder bed fusion of AlSi10Mg. Acta Materialia, 201, 231–243.

Mahmud, A., Huynh, T., Zhou, L., Hyer, H., Mehta, A., Imholte, D. D., ... & Sohn, Y. (2021). Mechanical behavior assessment of Ti-6Al-4V ELI alloy produced by laser powder bed fusion. Metals, 11(11), 1671.

Malý, M., Höller, C., Skalon, M., Meier, B., Koutný, D., Pichler, R., ... & Paloušek, D. (2019). Effect of process parameters and high-temperature preheating on residual stress and relative density of Ti6Al4V processed by selective laser melting. Materials, 12(6), 930.

Mooney, B., & Kourousis, K. I. (2020). A review of factors affecting the mechanical properties of maraging steel 300 fabricated via laser powder bed fusion. Metals, 10(9), 1273.

Mukherjee, T., Wei, H. L., De, A., & DebRoy, T. (2018). Heat and fluid flow in additive manufacturing–Part II: Powder bed fusion of stainless steel, and titanium, nickel and aluminum base alloys. Computational Materials Science, 150, 369–380.

Oyesola, M., Mpofu, K., Mathe, N., Fatoba, S., Hoosain, S., &Daniyan, I. (2021). Optimization of selective laser melting process parameters for surface quality performance of the fabricated Ti6Al4V. The International Journal of Advanced Manufacturing Technology, 114(5), 1585–1599.

Peng, T., Kellens, K., Tang, R., Chen, C., & Chen, G. (2018). Sustainability of additive manufacturing: An overview on its energy demand and environmental impact. Additive Manufacturing, 21, 694–704.

Pola, A., Battini, D., Tocci, M., Avanzini, A., Girelli, L., Petrogalli, C., & Gelfi, M. (2019). Evaluation on the fatigue behavior of sand-blasted AlSi10Mg obtained by DMLS. Frattura ed IntegritàStrutturale, 13(49), 775–790.

Poyraz, O., Ozkan, F., & Bas, O. O. (2021) Porosity in Ti6al4v alloy produced by powder bed fusion based additive manufacturing. Selcuk Summit 4th International Applied Sciences Congress, Karaman, Turkey, 24–33.

Prathyusha, A. L. R., & Babu, G. R. (2022). A review on additive manufacturing and topology optimization process for weight reduction studies in various industrial applications. Materials Today: Proceedings, 62(1), 109–117.

Rajabi, M., Simchi, A., Vahidi, M., & Davami, P. (2008). Effect of particle size on the microstructure of rapidly solidified Al–20Si–5Fe–2X (X= Cu, Ni, Cr) powder. Journal of Alloys and Compounds, 466(1–2), 111–118.

Rekedal, K., & Liu, D. (2015). Fatigue life of selective laser melted and hot isostatically pressed Ti-6Al-4V absent of surface machining. In 56th AIAA/ASCE/AHS/ASC Structures, Structural Dynamics, and Materials Conference (p. 0894).

Sagbas, B. (2020). Post-processing effects on surface properties of direct metal laser sintered AlSi10Mg parts. Metals and Materials International, 26(1), 143–153.

Sagbas, B., & Durakbasa, M. N. (2019, August). Industrial computed tomography for nondestructive inspection of additive manufactured parts. In Proceedings of the International Symposium for Production Research 2019 (pp. 481–490). Cham: Springer.

Sagbas, B., Gencelli, G., & Sever, A. (2021). Effect of process parameters on tribological properties of Ti6Al4V surfaces manufactured by selective laser melting. Journal of Materials Engineering and Performance, 30(7), 4966–4973.

Sames, W. J., List, F. A., Pannala, S., Dehoff, R. R., & Babu, S. S. (2016). The metallurgy and processing science of metal additive manufacturing. International Materials Reviews, 61(5), 315–360.

Slotwinski, J. A., Garboczi, E. J., & Hebenstreit, K. M. (2014). Porosity measurements and analysis for metal additive manufacturing process control. Journal of research of the National Institute of Standards and Technology, 119, 494.

Sreenivasan, R., Goel, A., & Bourell, D. L. (2010). Sustainability issues in laser-based additive manufacturing. Physics Procedia, 5, 81–90.

Srivatsan, T. S., & Sudarshan, T. S. (Eds.). (2015). Additive manufacturing: Innovations, advances, and applications. CRC Press.

Tang, M., Pistorius, P. C., & Beuth, J. L. (2017). Prediction of lack-of-fusion porosity for powder bed fusion. Additive Manufacturing, 14, 39–48.

Vilaro, T., Colin, C., & Bartout, J. D. (2011). As-fabricated and heat-treated microstructures of the Ti-6Al-4V alloy processed by selective laser melting. Metallurgical and Materials Transactions A, 42(10), 3190–3199.

Yang, T., Liu, T., Liao, W., MacDonald, E., Wei, H., Zhang, C., … & Zhang, K. (2020a). Laser powder bed fusion of AlSi10Mg: Influence of energy intensities on spatter and porosity evolution, microstructure and mechanical properties. Journal of Alloys and Compounds, 849, 156300.

Yang, T., Liu, T., Liao, W., MacDonald, E., Wei, H., Zhang, C., … & Zhang, K. (2020b). Laser powder bed fusion of AlSi10Mg: Influence of energy intensities on spatter and porosity evolution, microstructure and mechanical properties. Journal of Alloys and Compounds, 849, 156300.

Yasa, E., & Poyraz, O. (2020). Powder bed fusion additive manufacturing of Ni-based superalloys: Applications, characteristics, and limitations. In Additive manufacturing applications for metals and composites (pp. 249–270). IGI Global.

Yasa, E., & Poyraz, Ö. (2021). Investigation of residual stresses by micro indentation in selective laser melting.Lazerle metal tozergitmeprosesindekalintigerilmelerinmikrogirintitekniğiileincelenmesi. Journal of the Faculty of Engineering and Architecture of Gazi University, 36(2), 1029–1039.

Zhang, L. C., Attar, H., Calin, M., & Eckert, J. (2016). Review on manufacture by selective laser melting and properties of titanium based materials for biomedical applications. Materials Technology, 31(2), 66–76.

Zhu, H. H., Fuh, J. Y. H., & Lu, L. (2007). The influence of powder apparent density on the density in direct laser-sintered metallic parts. International Journal of Machine Tools and Manufacture, 47(2), 294–298.

Chapter 7

Spent Coffee Ground–Based Polymeric Materials for 3D Printing

S. L. Mak, W. F. Tang, C.H. Li, C.C. Lee, M.Y. Wu, W.Y. Chak, and W.K. Kwong

CONTENTS

7.1 INTRODUCTION

In recent years, coffee consumption has steadily increased, and experts predict that the global market will continue to expand. Coffee studies reveal fascinating information about consumer coffee usage, such as the type of people who drink coffee first thing in the morning or who enjoy taking coffee while still at work. According to assessments, about 30%–40% of the world's population drinks coffee on a daily basis. These percentages are substantially higher in the United States, accounting for almost 65% of the overall population. Despite the fact that coffee consumption is high in the United States, studies reveal that northern European countries are actual coffee addicts that consume massive volumes of coffee per capita. Coffee exports have amounted to 11.4 million

DOI: 10.1201/9781003306238-7

bags as at 2022 compared with the past years where they were lower. The total consumption rate has increased over the years and according to statistics of the fiscal year 2020/21, total 166,346 bags each of 60 kg were consumed.

The Scandinavian countries have the largest yearly per-person consumption, owing to their long, dark and harsh winters, which make coffee a cherished commodity. On per-person basis, the United States consumes around 4.4 kilograms (9.7 pounds) of coffee per year, ranking it as the world's 25th largest consumer. In the United States, a regular person consumes three cups of coffee every day. Conferring to a National Coffee Association survey, Hispanic Americans (65%) are further likely than other races to drink coffee every day in the United States. Coffee is consumed on a daily basis by 64% of Caucasians. Coffee is consumed on a daily basis by 54% of individuals who identify as African Americans (Food Truck Empire, 2021). It is predicted that global coffee consumption will grow by a third by 2030, adding 200 million bags to the total. The reasons for this are wage growth and population growth.

Waste generation is a natural occurrence, and the amount of waste created is directly related to cultural and lifestyle changes. Coffee grounds are a one-time-use product, and the total waste generated by coffee disposal equals all imports and sale of coffee combined. Coffee has massive environmental consequences, generating massive amounts of solid and liquid waste all over the world (Pujol et al., 2013). This is owing to a significant shift in cultivation practices. Coffee is grown organically under a shaded canopy of trees in tropical and subtropical countries at high elevations, providing a cherished habitat for native animals and creepy-crawlies, as well as reducing topsoil corrosion and eliminating the need for chemical fertilizers.

As a result of separating the commercial product (the beans) from its source (the cherries), a large amount of pulp, residual residue and parchment is generated. Waste coffee grounds are an underutilized high-nutrient material with promise as horticulture compost, thus recycling by vermicomposting can be a low-cost, long-term alternative to disposal. In a six-month period, it is projected that processing 547,000 tons of coffee in Central America produces 1.1 million tons of pulp and pollutes 110,000 cubic meters of water (Fernandes et al, 2017). Coffee pulp is frequently dumped into streams, drastically harming sensitive ecosystems. This extra waste can also cause havoc with soil and water sources. If not properly disposed of, used coffee grounds are considered a possible harm to the environment. Used coffee grounds can leak methane into the atmosphere, contributing to the global climate change problem.

7.2 CURRENT APPLICATIONS OF SPENT COFFEE GROUND

7.2.1 Fertilizers

Spent coffee grounds are derivatives from the coffee industry, which are rich in carbon content and could be utilized in many applications in different sectors. Anecdotal evidence suggests that locally produced discarded coffee grounds can be used in town cultivation and botanical gardens, either directly in the soil or after composting with other urban carbon-based wastes. Coffee grounds are a great source of nitrogen for compost pile. Adding organic material to the soil using coffee grounds as a fertilizer improves drainage, water holding and ventilation. The spent coffee grounds also charm earthworms and help bacteria that are beneficial to plant growth flourish.

While it may appear that utilizing free coffee grounds is the ideal approach, several gardeners have discovered that putting coffee grounds directly on the soil has had terrible results for their plants. The grounds can be applied in different ways which are (1) sprinkling the coffee grounds directly into the garden, (2) mixing of the coffee grounds into compost, which boosts the nitrogen content and (3) steeping of coffee grounds in a bucket of water overnight, which makes a liquid fertilizer.

7.2.2 Biofuels

Biodiesel is a renewable energy source made from natural oils and fats that is used to replace petroleum fuel in vehicles. Domestically, this clean renewable fuel can be made from a range of oils, fats and greases using chemical procedures. An ASTM definition for biodiesel is a blend of long-chain monoalkylic esters derived from fatty acids obtained from renewable sources that is suitable for use in diesel engines. This latter not only meets the majority of the standard features of petro-diesel, but it also offers a number of traits that make it a very promising alternative energy source (Atabani et al., 2019). Because SCG contains a high amount of organic substances like as fatty acids, lignin, cellulose, hemicellulose and other polysaccharides, they have a lot of potential as a biofuel feedstock.

Contemporary research studies have highlighted the value of SCG as a valued spring of phenolic composites and bioenergy. SCG has been assimilated in biogas production and bioethanol production, biodiesel and bioethanol have been produced from SCG, coffee oil has been extracted from SCG using four solvents and prototype scale extraction and biochar has been produced from defatted SCG using slow pyrolysis. Because spent coffee grounds (SCG) have a high calorific value, they can be used to make refuse-derived fuel (RDF). The disadvantage of pure SCG pellets is that they may reduce boiler performance and cause particles to be released, so more materials are needed to produce good quality pellets. Past researches reaffirm the use of waste paper and coffee remnants for briquette manufacturing, the manufacturing of carbonized lump charcoal from Rain tree (SamaneaSaman) and SCG/tea waste, the impact of blending SCG and coffee silverskin (CS) on the value of pellet fuel produced and the manufacturing of eco-fuel briquettes with 32% spent coffee ground, 23% coal fines and 11% sawdust to profit lower harmful emissions compared to fossil fuels. (Patcharee, 2015).

In the last decade, much research has been done to burn biomass waste in various boilers, for example, for the assessment of the pyrolysis tests in a commercial residential wood pellet boiler with a pure SCG pellet, a blended pellet (50% SCG and 50% sawdust) and a sheer pine wood pellet, for the fuel and combustion test in a small boiler (6.5 kW) with SCG, and for the ignition tests of wood pellet on a fixed bed reactor with various co-products. The burning of straw, olives, tomatoes, cocoa beans and other similar materials resulted in relatively good tank productivity, but the delinquent is the resulting ash, which has a truncated melting point. It is vital to keep the heat exchanger from becoming clogged. It is far more advantageous to mix these waste biofuels with wood, so eliminating this unpleasant issue. We would be able to cut waste while also obtaining a green energy source by burning this biofuel.

7.2.3 Biodiesels

An in situ trans esterification process could be used to convert the lipids in SCG to biodiesel. Existing in situ trans esterification of wet SCG biomass is energy costly, as it is carried out at a high response temperature to minimize the water outcome and response time. To generate biodiesel straight from wet SCG biomass, a new method was devised that includes simultaneous extraction and trans esterification in a single step utilizing a soxhlet device. With hexane as a co-solvent, a homogeneous base catalyst at 0.75 M demonstrated greater catalytic action than acid on fatty acid (FA) abstraction efficacy and FA to fatty acid methyl ester (FAME) alteration efficacy (Abdullah et al., 2017). The maximum FA to FAME alteration proficiency of 97% was achieved by studying the factorial result of methanol to hexane ratio and response time at a ratio of 1: 2 and a reaction period of 30 minutes. Furthermore, the reagent could be reused five times before dropping its effectiveness. When compared to conventional technologies, the sensitive abstraction soxhlet (RES) approach might save 38%–99% on energy consumption.

SCG (spent coffee ground) has been shown to contain lipids in the 10%–28% range. Some researchers have confirmed that in situ trans esterification of crop seed/biomass can create a high biodiesel production. In addition, biodiesel produced using the in situ trans esterification technique passes ASTM standards and can be utilized to control diesel engines. Biodiesel manufacturing now depend on vegetable oil as a spring, exacerbating the food vs. fuel dilemma. The use of feedstocks that need a lot of water to cultivate (drink vs. fuel) has been criticized as well. As a result, finding low-cost feedstock's with high lipid content is important to improve biodiesel's competitiveness. Despite the fact that unused cooking oil is two to three times cheaper than vegetable oil, sources are restricted and rich in contaminants.

7.2.4 Bricks

Because of the high fire temperature and greenhouse gas emissions, the typical firing process for manufacturing construction bricks is not sustainable. As a result, there is a rising incentive to invest in environmentally friendly technologies. SCGs are solid wastes produced by coffee consumption. Every day, six million tons of coffee grounds are created around the globe. As a result, enormous amounts of SCG are burned and buried, potentially releasing hazardous compounds like polyphenol and tannin into the environment. SCG recovery, reuse and potential in the engineering area are all possibilities because to their intriguing physical and chemical properties.

Clay bricks are made through thermal behaviour to get a better enclosure lagging when made from the grounds. During the fire process, organic compounds combust within the matrix, resulting in increased porosity. As a result, the heat conductivity of bricks is lowered. However, other characteristics that impact the utility of the bricks, primarily the percentage of water immersion and compressive power, are influenced by porosity. Several experiments were carried out for various percentages of trash in order to generate bricks that met governing criteria, were established by standards and had the lowest thermal conductivity possible (SpringerLink, 2015). The findings were examined and compared to earlier studies, finishing that it is probable to add 17% waste while the compressive power of the bricks is greater than 10 N/mm2, indicating that they can be used for physical reasons. However, because bricks manufactured by adding 17% must be coated, they cannot be utilized as facing bricks. Thermal conductivity is lowered by up to 50% in this situation (Figures 7.1–7.3).

7.3 COMPARISON OF DIFFERENT TREATMENTS (OIL REMOVAL PROCESS)

7.3.1 N-Hexane extraction

Hydraulic pressing, expeller pressing and solvent extraction have all been used in the past to excerpt oil from oilseeds (SE). Solvent extraction is one of these technologies that has been widely used for cost and practical reasons. Oilseeds are treated prior to solvent extraction in order to maximize oil recovery. The SE process involves washing the oilseeds with hexane, and then separating the hexane from the oil by evaporation and distillation. Because of its rapid oil recovery, narrow boiling point (63–69 °C) and outstanding solubilizing ability, hexane has been commonly employed in the SE process (Liu and Mamidipally, 2005). N-Hexane extraction is time consuming since it requires large amount of a solvent (Hexane) and energy, which results in low productivity.

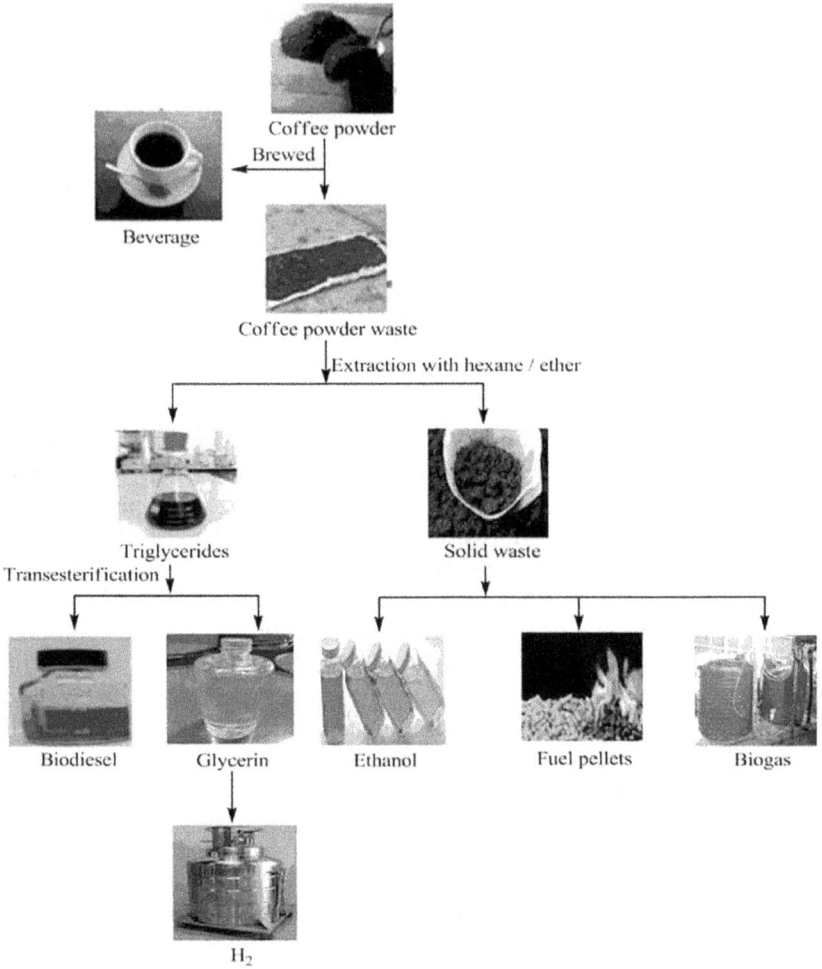

Figure 7.1 Flow of converting spent coffee grounds (SCG) to Biodiesel and Solid Biofuel.

Inhalation of hexane-contaminated air in a short-term affects the nervous system causing dizziness, nausea, headaches and even coma. Chronic contact can harm the neurological system more severely. If swallowed, it can cause significant abdominal pain and have a negative effect on the respiratory system, causing shortness of breath, coughing, mouth, throat or chest burning, and possibly chemical pneumonitis. When working with hexane, wearing personal protective equipment is recommended. Hexane should be stored in fireproof containers in a well-ventilated environment, away from strong oxidants and in well-sealed containers. Any unused chemical should be recycled or returned to the supplier for its intended purpose.

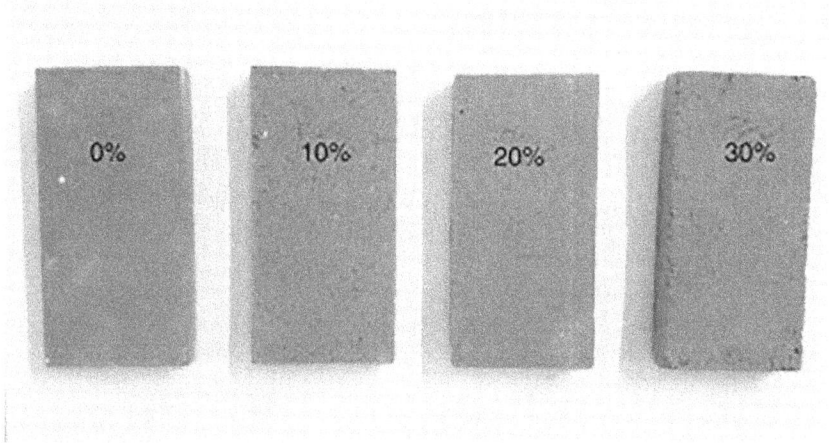

Figure 7.2 Bricks made of Moroccan clay with different percentage of spent coffee grounds (SCG).

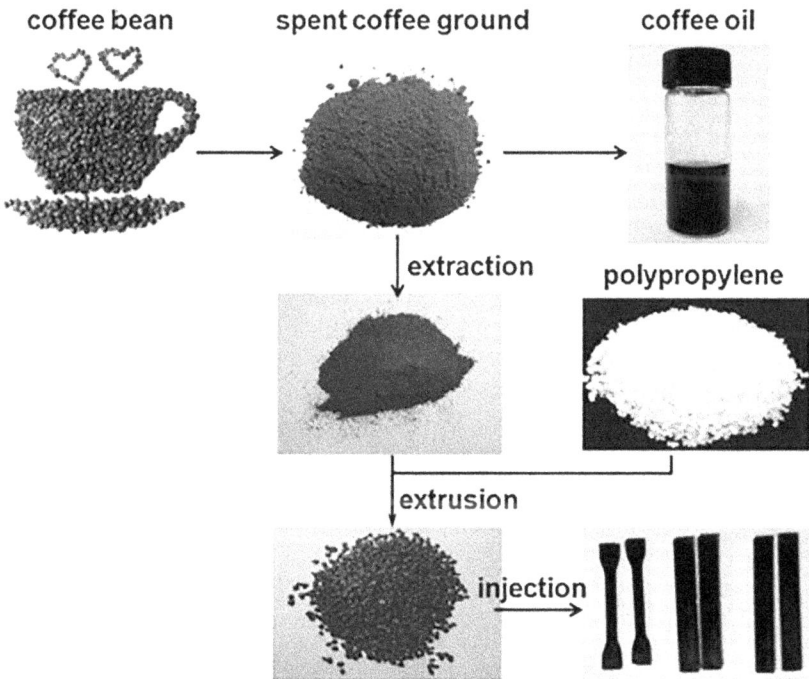

Figure 7.3 Process Flow of making products by using SCG-PP composite materials.

7.3.2 Ultrasonic extraction

Ultrasonic extraction is a frequently used method for extracting plant materials using fluid diluters this is because the superficial area between the solid and liquid segments is considerably greater due to cell disturbance and particle dispersal. It has been shown to be a faster and more complete abstraction practice than traditional methods. When using traditional extraction procedures like liquid-liquid extraction in solvents, the extraction efficiency generally improves as the temperature rises. Because temperature influences the stability of the phenolic components, this frequently results in extract degradation and loss of quality. The ultrasound-assisted solid-liquid extraction method has been found to be both successful and time-saving. Because the ultrasonic pressures are so intense, they provide all of the energy required for extraction, requiring fewer or even no solvents. Because the sonicated batch or flow cell reactor can be efficiently cooled, the temperature can be well controlled (or heated if necessary). Hielscher Ultrasonics also offers ATEX and FM certified explosion-proof ultrasonic equipment for solvent extraction procedures.

The spent coffee ground is still a raw material rich in extractable components thanks to the powerful extraction pressures of ultrasound. Coffee trash is a suitable raw material for extracting the residual active chemicals because it is inexpensive and readily available. Although the caffeine and other components in coffee trash are lower than in wasted coffee powder, there is still a significant amount that can be extracted. The full influence on the processing conditions becomes especially important when it comes to releasing these chemicals from the coffee ground. Ultrasound with high power can extract large amounts of active chemical in a short length of time.

7.4 MIXING SPENT COFFEE GROUND TO RECYCLED POLYMERS

The moisture content of the coffee by-products is removed and dehydrated to underneath the orientation moistness level, then extruded and pulverized at a high temperature. Coffee by-products are cooked at a temperature of 50 °C to 120 °C for 60 to 240 minutes. The water is dried to a weight of 6%–7.5% or less. The coffee by-products are chilled after initial grinding and then crushed to a size of 120 to 300 meshes using a crusher according to the presentation. When the particle size of coffee by-products is smaller than 120 mesh, they become so large that the surface of the bio plastic product is rough, affecting the strength and elongation of the final product. The standard has deteriorated and the physical qualities and colour variation of the fibre by-products of the coffee by-products are destroyed, the smell is diminished and the crushing takes a lot of time and exertion to lower efficiency if the particle size surpasses 300 meshes. As a result, it was determined that the coffee by-grinding product's particle size should be between 120 and 300 mesh.

To make a bio plastic pellet, principal raw extrude is blended with polypropylene, polyethylene, or a combination of the two and extruded (130). This phase eliminates the volatile gas from the kneader by heat and reflux, then plasticizes and thaws it at 180 °C to better spread and squeeze the uncooked materials. A twin screw extruder produces a pellet with a width of roughly 5 mm and a size of about 2 to 3 mm at a temperature of 150 to 220 °C. Polypropylene or polyethylene, or a combination of the two, is further added as a biochemical raw material in amounts ranging from 47% to 65% by mass, depending on a final product weight of 100%. More particularly, polypropylene, polyethylene, or a combination thereof helps as a folder to increase the malleability of the current creation's bio plastic, and when the synthetic resin content is less than the reference value, the bio plastic's elongation and tensile strength become weak. Natural coffee by-products' aroma or colour is limited above this threshold, interfering with biodegradation. Polypropylene and polyethylene can be employed separately or in combination, with the fraternization ratio adjusted according to the physical qualities necessary for the present invention's bio plastic, a coffee waste product. Finally, bio plastic pellets are injected to manufacture containers and the like, and extruded to make films and the like. The injection and extrusion procedures are identical to those used for recognized plastics, and current equipment is used.

SCG polymers have covalent bonds in matrices like polyethylene (PE), polypropylene (PP), PU, Poly(Lactic Acid) (PLA) and Poly(Butylene Adipate-Co-Terephthalate) (PBAT). Polyethylene (PE) compounds are the most commonly used artificial polymer milieus filled with SCG for their ubiquitous use in wrapping, truncated costs and generally good qualities. Many varieties of polyethylene exist, among them are high-density polyethylene (HDPE) and low-density polyethylene (LDPE). PE commonly used thermo-plastic poly-olefin for blast and injection moulding in the world. PE is utilized in a variety of applications, including tubes, panes, containers and other similar goods, due to its great durability, easiness of production, little electrical conductivity and bio-chemical motionlessness. In comparison to LDPE, HDPE has superior impression strong point and electrical lagging. Polypropylene (PP) compounds, also known as polypropylene, are a useful commodity thermoplastic that is well-known for its low price and wide range of uses, including film, fibre, packaging and automotive parts (Hatakeyama and Hatakeyama 2009).

PP can be recycled unaided or as a conditions for amalgams containing a variety of reinforcements (particles, fibres etc.), including bio-based materials like wood. PU Composites – used to replace rubber – are made by combining an isocyanate and polyol to form a polymer. Insulators, rigid foams, coatings, adhesives and elastomers are some of the uses of PU composites. Efforts are focused to generating polyol from an ordinary source, as well as

Table 7.1 Different mechanical and physical properties of SCG composites according to PP

Formulation (wt. %)	Neat PP	PP/CC 75/25	PP/SCG 75/25
Tensile Yield Strength (MPa)	34.8 ± 0.1	28.2 ± 0.2	24.2 ± 0.4
Tensile Modulus (MPa)	1650 ± 10	2200 ± 10	1530 ± 20
Elongation at Yield (%)	7.5 ± 0.1	3.3 ± 0.1	4.8 ± 0.1
Elongation at Break (%)	·700	5.1 ± 0.3	7.9 ± 0.9
Maximum Flexural Stress (MPa)	55.0 ± 0.1	52.5 ± 0.7	43.3 ± 0.5
Flexural Modulus (MPa)	1930 ± 10	2510 ± 50	1650 ± 30
HDT (°C)	110.9 ± 0.4	126.9 ± 1.4	102.6 ± 3.1
Izod Impact Strength (J/m)	22.4 ± 0.2	24.8 ± 1.0	24.0 ± 2.1
MFI (g.10 min^{-1})	10.1 ± 0.1	7.6 ± 0.1	9.9 ± 0.1
Density (kg/m^3)	0.910 ± 0.004	0.995 ± 0.004	0.985 ± 0.007

PU compounds with natural or waste-based plasters, due to environmental concerns. So far, only a few studies have been published on the use of SCG as filler in PU compounds. Poly(Lactic Acid) (PLA) composites bio-degradable polymers are said to be extra expensive. As a result, PLA composites including lignocellulose trashes have been developed to lessen PLA content, making them more inexpensive for use as a replacement for synthetic polymers in various applications. Several studies have described on SCG valorization as a plaster for PLA compounds to lessen the quantity of SCG in the surroundings over the last ten years. Researchers explored the coupling effect of SCG in PLAs and 4,4-methylene diphenyl diisocyanate (MDI) as a coupling agent (10, 20, 30 and 40 wt.%) (Hatakeyama and Hatakeyama, 2009). The mechanical strength of the PLA and SCG dropped as the filler content increased; however, MDI was demonstrated to advance the morals at 30 wt.% by forming a urethane link amongst them. Poly(Butylene Adipate-Co-Terephthalate) (PBAT) composites are bio-degradable polymers commonly used in packaging and biomedicine. PBAT was employed as a medium for lignocellulosic compounds for the reason that of ithas a low thermo mechanical characteristics and high manufacture cost (Hatakeyama and Hatakeyama, 2009). Coffee waste is currently being used in PBAT compounds with coffee husks and silver skins. Table 7.1 shows the different mechanical and physical properties of SCG composites according to PP.

7.5 BENEFITS OF THE SCG POLYMERS

Polyphenolics, lipids and lignin make up the majority of SCG. The abundance of polyphenolics such as chlorogenic acid, caffeic acid, gallic acid, ferulic acid and cinnamic acid makes it possible to create high-value-added yields, particularly in the cosmetics and skin care industries. However, in order to retrieve useful molecules from SCG, an extraction procedure is required.

Unique Polymer Systems offers a variety of solvent-free polymer coatings for industrial pump repair that have a number of advantages like longer pump life and improved efficiency, reduced energy costs, extreme abrasion resistance, increased chemical and temperature resistance, chemical adhesion, which eliminates the need for hazardous solvents, versatility in application on various substrates and higher hardness than other coating products. Unique Polymer Systems has established itself as a world leader in this industry thanks to these coatings. To generate coatings with a high level of corrosion and erosion resistance, products have been developed for nearly any application. They are also physically and mechanically strong enough to resist the harshest industrial settings.

SCG polymer provides solutions that reduce resource usage while increasing resource efficiency. This high-quality plastic resin manufacturing process produces a unique HDPE that strikes the perfect combination of strength and stiffness, requiring less material to build a product with the same level of strength. From lightweight CSD caps and closures to high-impact films for industrial use, to SCG Green Choice-certified extra-strong big chemical tanks, HDPE lends itself to a wide range of products. More importantly, when compared to standard plastic resins, SCG polymer reduces greenhouse gas emissions during the manufacturing process. SCG polymer offers solutions that can convert multi-material packaging, which is widespread but notoriously difficult to recycle, into mono-material PE or PP packaging that can be recycled economically and has functional and aesthetic features that suit the objectives of brand owners.

Sorted and cleaned plastics are processed into high-quality PCR resins that not only suit user needs but also assist brand owners in achieving their goal of employing recycled materials. These PCR resins are also certified to the Global Recycled Standard (GRS), which verifies the source of recycled plastic resins' basic components. SCG Chemicals has cooperated with both local and worldwide partners with expertise in recycling, such as Teamplas, to develop and grow recovered plastic markets to other locations. Plastics that have not been fully separated can be recycled as feedstock for the creation of plastic resins with qualities equal to virgin plastics that have achieved ISCC PLUS accreditation through a method known as advanced recycling, sometimes known as chemicals recycling.

SCG polymer also provides biodegradable plastic options. Chemicals' special formulation allows them to be easily extruded into household and industrial film products, boosting convenience and moulding efficiency. They also feature a wide range of properties that fulfil user needs and have been certified as bio-compostable by the world's premier German-based institution DIN CERTCO. Another option for minimizing the use of rapidly diminishing fossil-based resources in plastic manufacture is to switch to resources that can be replaced through cultivation, such as bio-based resources, which emit fewer greenhouse gases and so assist to reduce global warming's impact.

7.6 APPLICATIONS OF SCG POLYMERS

7.6.1 Injection moulding

Injection moulding (or molding, as it is spelled in the United States) is one of the most important processes for producing plastics articles, along with extrusion. Whether it's high-precision technical components or disposable consumer goods, it's a quick and reliable method for mass-producing identical objects. Developments like the OVDesigns have emerged with an Espresso polymer, which is highly suitable for products with log lifespan. The material is 100% recyclable and compostable in an industrial setting. The product can either be recycled into a new product or decomposed in an industrial composting facility. Coffee grinds and sugarcanes make up the Espresso polymer. This well-formulated combination produces a strong material with wall thicknesses ranging from 1 to 10 mm. Any injection moulding, rotation moulding, or extrusion business can process the material. Because the material can be used in existing machinery, there is no need to invest in new machinery because the investment will be comparable to that made with conventional polymers such as ABS, PE, PS or PP.

7.6.2 Compression moulding

According to the current invention, this is a method of making moulded goods from coffee grounds that allows for little chemical addition during moulding, resulting in practice materials that are safe for human consumption. The process entails use of high density polythene (95%) weight and coffee husks to reinforce the filler (5%) weight. The husks are disposed to open air burning and are used to help reduce their negative impacts on the environment. A moulding machine is made of mild steel and stainless steel consisting of heating chambers, mould base, compression shaft and observation windows. The temperature of the heating chamber is controlled using a temperature controller. The created bio-composite polymers are put through elongation, tensile strength and water absorption tests. Coffee husks reduce the tensile strength and percentage elongation when used in compression moulding.

7.6.3 3D Printing

Composite mixture of PLA and coffee ground have been developed, which are tough and environmental friendly. The grounds help reduce the manufacturing costs while still maintaining abetter recycling. The PLA composite isn't made from actual coffee grounds. It entails the use of a dry, odourless material that is left over after the coffee oil and bio-diesel have been removed. When 20% of these leftovers are mixed into ordinary PLA, the material gains a 400% increase in toughness over standard PLA. Due to the additions, the material has a highly earthy colour and can be created with a standard filament desktop 3D printer. Another benefit of this new composite is that it may be less expensive than pure PLA. The waste's inclusion lowers the overall material cost while increasing the volume of the overall material.

7.7 CONCLUSION

The consumption rate of coffee was increased to release much spent coffee ground and pollute the earth. It is a room for the researchers to discover how to properly use the waste materials. By extracting the oil from the SCG, the oil can be used a biofuel and lubricants. And the SCG is mixed with polymeric wastes, the green composite materials is able to increase the tensile strength.

ACKNOWLEDGMENT

The work described in this paper was substantially supported by a grant from the Research Grants Council of the Hong Kong Special Administrative Region, China (UGC/FDS16/E01/20).

REFERENCES

Abdullah, S. H. Y. S., Hanapi, N. H. M., Azid, A., Umar, R., Juahir, H., Khatoon, H., & Endut, A. (2017). *Renewable Sustainable Energy Reviews*.
Atabani, A. E., Ala'a, H., Kumar, G., Saratale, G. D., Aslam, M., Khan, H. A., Said, Z., & Mahmoud, E. (2019). Valorization of spent coffee grounds into biofuels and value-added products: Pathway towards integrated bio-refinery. *Fuel*, 254, 115640.
Eco-fired clay bricks made by adding spent coffee grounds: A sustainable way to improve buildings insulation. (2015, January 14). SpringerLink. https://link. springer.com/article/10.1617/s11527-015-0525-6
Fernandes, A. S., Mello, F. V. C., Thode Filho, S., Carpes, R. M., Honório, J. G., Marques, M. R. C., … & Ferraz, E. R. A. (2017). Impacts of discarded coffee waste on human and environmental health. *Ecotoxicology and Environmental Safety*, 141, 30–36.
Hatakeyama, H., & Hatakeyama, T. (2009). Lignin structure, properties, and applications. In *Biopolymers* (pp. 1–63). Springer, Berlin, Heidelberg.
International Coffee Organization – What's New. https://www.ico.org/prices/new-consumption-table.pdf

KR101344471B1 – Bio plastic using coffee residual products and method making the same – Google patents. (2013, December 24). Google Patents. https://patents.google.com/patent/KR101344471B1/en

59 global coffee industry statistics and consumption trends – Updated for 2021. (2021, March 16). Food Truck Empire. https://foodtruckempire.com/coffee/industry-statistics/

Liu, S. X., & Mamidipally, P. K. (2005). Quality comparison of rice bran oil extracted with d-limonene and hexane. *Cereal Chemistry*, 82(2), 209–215.

Patcharee, P., & Naruephat, T. (2015). A study on how to utilize waste paper and coffee residue for briquettes production. *International Journal of Environmental Science and Technology*, 6, 201.

Pujol, D., Liu, C., Gominho, J., Olivella, M. À., Fiol, N., Villaescusa, I., & Pereira, H. (2013). The chemical composition of exhausted coffee waste. *Industrial Crops and Products*, 50, 423–429.

Chapter 8

Recycling Polyethylene Terephthalate to Make 3D Printing Filaments

S. L. Mak, W.F. Tang, C.H. Li, C.C. Lee, M.Y. Wu, W. Y. Chak, and W. K. Kwong

CONTENTS

8.1 INTRODUCTION

China processed around 67% of the global plastic waste in 2016. From 2017, the Chinese government stopped importing the industrial and domestic waste from other countries. Jenna Jambeck (University of Georgia) indicated that

DOI: 10.1201/9781003306238-8

this policy could lead to disposing of around 111 million metric tons of plastic waste to landfill by 2030. Around five million plastic drinking bottles are disposed in Hong Kong daily. The bottles are made of thermoplastics, such as polyethylene (PE), polypropylene (PP) and polyethylene terephthalate (PET) materials. There is a lack of sufficient facilities to process and recycle the waste bottles.

Three-dimensional (3D) printing was developed since the 1980s and has seen significant growth in the past 10 years. Three-dimensional printing is one of key technologies in the framework of Industry 4.0. The material extrusion was applied to develop the Fused Filament Fabrication (FFF) or Fused Deposition Modelling (FDM), which is now available to domestic users. The polymeric material such as polylactic acid or polylactide (PLA) is widely used to make the filaments for making the product. The polylactic acid material is a biodegradable material derived from plants such as corn starch, cassava roots, chips or starch. Due to low cost of FFF machine and filament material, the waste materials has increased at a high speed.

Earlier 3D printing filaments were used to be extruded, while existing extruding machine use a screw to apply the elevated temperature and high pressure to melt the polymeric material and push the material through the small die hole to make the filaments. The limitations of single screw extrusion include (1) poor mixing capability for different recycled polymeric materials and (2) melted polymer may block extrusion machines and give poor printing quality. As single-use PET drinking bottles was difficult to be recycled in Asia, it is a possibility for the industrialists and scientists to evaluate how the materials can be used for 3D printing.

8.2 GLOBAL CONSUMPTION RATES OF PET POLYMERS IN THE PAST 10 YEARS

Global consumption trends for PET polymers are massively increasing. The general plastic consumption has been high but the PET polymers' popularity has been confirmed by the recyclability. Countries are tired of plastic pollution and require polymers with high recyclability. In 2012, Asia-Pacific projected consumption trend stood at 29.4%, while the US and Western Europe witnessed 24.1 and 19.7%, respectively (BNP Media, 2012). With market projected to grow at 5.2% annually is a massive increment in the use of PET packaging. From a few tones in 1950s to over 15 million tons in 2012 and close to a billion tons in 2021, the PET consumption trend is alarming.

8.3 CURRENT COMMON APPLICATIONS OF PET POLYMERS SUCH AS MAKING DISPOSABLE DRINKING BOTTLES

Three-dimensional printing is the process of producing plastic that could not have been possible through molding. Traditionally, plastics were made to appear into different shapes such as a bottle, plate or bowl by molding. The bottles and plates are considered to be two-dimensional (2D) shapes. However, there can be more shapes that are difficult produce just by molding a plastic. This is where 3D comes in. Currently, PET can be considered very popular in the 3D printing business because it is being used by many manufacturers (Bainbridge et al., 2020). Resin 3D printing is the technology in which liquid resin is stored in a container instead of being injected through a nozzle. The major terms or types of resin 3D printing include stereolithography (SLA), digital light processing and liquid crystal display (LCD) or masked stereolithography (MSLA) (Zander et al., 2019). PET serves the printing world in many ways. It can be the container that holds the liquid used in printing. It can also be the raw material used to print out different shapes. Using it as a raw material works because of PET's chemical properties. For example, in the semi-liquid shape, it can be spread on the top of a plate and the printer passed over it. The printer is able to mold into any shape because of the adhesive forces that pull it apart instead of together. In this case, it does stick too tight, enabling manufacturers to print any shape they want. Its filament prints easily and produces an excellent layer adhesion. It does not shrink rampantly, which makes it ideal for large prints. PET does not produce any odor when being printed, and is strong and chemical resistant.

It is used to make disposable drinking bottles. The ability to make good food and drink containers emanates from the fact that PET/PETG can wage a good chemical resistance and can be thermoformed without difficulty. Therefore, it has been widely adopted to develop oil containers, drinking containers and food storage boxes. These packaging materials are all FDA-compliant since they do not pollute the environment when disposed of since PET is highly biodegradable. Cosmetics consumers are also likely to prefer PET because of its light weight and strength, which makes it easier to distribute.

It is applied in medical and pharmaceutical applications. Medical equipment require tough materials that cannot break easily or get damaged during sterilization processes. PET's rigid structure makes it a good fit for the application in the health sector (Bainbridge et al., 2020). It can withstand these harsh processes. It also has wide applications in the making of medical implants, which can mostly be credited to its biodegradable property. In addition to the medical devices, PET is still being used to make packaging materials for the devices, making it one of the most popular polymers in the health provision industry.

PET can also be seen at retail stands and displays since it does not break easily. However, it best fits this role because of the ease with which it can be colored. It is equally ideal for signage (Zander et al., 2019). Naturally transparent materials can be colored without problems, and since PET falls in this category manufacturers have been able to apply different colors to it to create different appearances. Since it can easily blend, many colors can be applied, which makes it ideal for signage and displays.

PET has been applied as machine guards. It forms plastic covering that provides protection to users. These guards have been effectively applied in the food processing units to protect workers, their formation takes less time than polycarbonates and these are more durable than acrylic.

8.4 WAYS TO HANDLE PET WASTES

There is no better way of handling plastic wastes than trying to minimize the waste. People ought to avoid single-use plastics (Gopinath et al., 2020). These plastics are normally manufactured to serve one purpose only. For example, a drinking straw is manufactured to help beverage consumers draw drinks from bottles. After the consumer has finished the drinks from a bottle, he throws away the straw. If the person were to consume three other bottles, four straws will be in the dustbin just from one person. Another example is plates, which are known for single uses. Some plates are used to serve only one type of food, such as chips after which they cannot be used for other purposes. There are numerous occasions where plates are discarded not because they are torn or damaged but because people have used them in one meal or the other. These plates should not be thrown away after a single meal. If consumers cannot put these plastics into numerous uses, they need to adopt their alternatives.

PET polymers can make good bags for those who love shopping. However, some people still make these bags single-use materials. It is advisable that people uncomfortable with plastics should use cloth bags because it will help reduce the level of plastic consumption. However, those who use plastic bags should know that PET polymers are biodegradable (Gopinath et al., 2020). However, they are still plastics and require close attention. For example, a plastic bag can be used more than once to avoid creating a lot of waste product within a short time. Therefore, whenever an individual thinks of filling up his fridge, which normally occurs once a week, he should leave the house with at least one reusable bag. This will minimize what he can call as plastic waste in his household because he is reusing most of it.

People should avoid careless disposal of plastics that they cannot use again. For example, it is next to impossible to try and use chewing gum again after consuming it. It is noteworthy that chewing gums are made of plastics. Chewing gums can be recycled. However, most people choose to throw it away once they are done with its taste. Since it is non-biodegradable, the gum will not be disappearing any time soon after it is thrown into a garden, lawn or by the roadside. There are correct recycling containers where chewing gums could be placed. However, the best way is to reduce consumption by going for alternatives such as the natural and organic chewing gums.

People should buy more bulk food and fewer packaged products. Disposable containers are inundating supermarkets. For example, any visit to a supermarket or a shopping mall will point an individual towards the direction of PET bottles. Some of the things packaged in the PET bottles cannot be repackaged elsewhere, but the consumer can buy less of products put in plastic bottles and more of the alternatives (Khoo et al., 2021). If the consumer takes the PET bottles, he should think about recycling and a safer disposal. Buying in bulk has been a better way of avoiding excessive plastic consumption.

Avoid comingling the plastic with other waste products. Since people have busy lifestyles; they can forget to pay attention to how they place the plastic after use. For example, they can throw it in the wrong container where it mixes up with other items. Plastics can release harmful materials into wrong places within the house. It should be handled carefully, especially by avoiding the mix up with other waste products.

It can do good to teach one's family about the importance of handling plastic properly. This step can be ignored by many, but it helps to create awareness about the handling of plastic waste. People tend not to think about the harm the plastics cause to the environment, and even less about the harm they can cause to human beings. Therefore, it pays to inform family members that they should be careful around plastics.

8.5 DISPOSAL

Plastic's best disposal method is that which takes it from the consumer to the landfill where the municipal authorities can flag it for recycling. The disposal process begins at the consumer's home. After consuming the contents from the plastic, such as drinking water from a PET bottle, the consumer should place it safely in a dustbin (Tulashie et al., 2019). These bins are tucked outside the house in the developed nations such as the US. Putting it outside makes it easy for the municipal authorities to collect even if the homestead owner is not available. The consumer should ensure that the dustbin for plastics is not comingled with other waste materials. Every time the consumer empties contents from a PET bottle he adds the bottle to the dustbin's contents, presumably other plastic containers.

After the plastic containers have stayed for a certain period, the household collection is handed over to the municipal collectors. The household collection occurs individually most of the time, but there can also be a central point where all the waste from a neighborhood is collected (Tulashie et al., 2019). At the moment, individual houses are provided with litter bins near their houses where they put in the used plastic materials. The authorities take the collected plastic waste to the landfills. These refer to large holes created as a central point for the collection of all plastics used by the municipality across the city or town. From the landfills, the plastics are taken to the recycling points. Plastic waste at the landfills is referred to as post-consumer municipal plastic waste. The municipal authorities can provide licenses to different companies to recycle the gathered waste so long as such recycling processes do not harm the environment farther.

8.6 RECYCLING METHOD AND APPLICATIONS OF RECYCLED PET

Recycling starts with sorting. The used PET plastic materials are taken to the materials recovery facilities (MRF) where they are sorted to separate them from waste from other materials. For example, they can be sorted to separate metals. Ballistic sorting can also remove dust or films. Spectral sorting uses sensors to detect polymer type and color. The waste bottles are then flattened and put in bales for shipment to processing centers.

Companies mostly use mechanical recycling. It involves melting the resin, filtering and extruding or molding it into new PET articles, such as bottles, films, strapping or fibers. Sometimes the PET feedstock can fail to produce pure enough products for mechanical recycling and sometimes manufacturers fail to chemically recycle the resin back to monomers or oligomers. Reaction products such as terephthalic acid (PTA) or dimethyl terephthalate (DMT) and ethylene glycol (EG), or bis(2-hydroxyethyl) terephthalate (BHET) can be used.

8.6.1 Mechanical recycling

The bottle flakes can be re-pelletized. The bottles are transformed into flakes. They are then dried and crystalized. The granules look clean because they have been filtered and plasticized. The end product reveals an amorphous re-granulate whose viscosity ranges from 0.55 to 0.7 (Gopinath et al., 2020). The viscosity depends upon pre-drying of the flakes. Since the acetaldehyde and oligomers are present in lower levels, the viscosity reduces.

8.6.2 Converting to bottle flakes directly

The PET flakes can be turned into usable bottles with the help of polyester intermediates. They can be produced through strapping mills, which are effective at adjusting the viscosity to the required level. Reconstituting the viscosity is possible because of polycondensation. With adequate viscosity, the granules can stick together through spinning or any other method.

8.6.3 Chemical recycling

There are several ways to recycle the PET chemically. One is through partial glycolysis. It enables the polymerization of the PET granules by converting the rigid polymer into short-chained oligomers which can be melt-filtered at low temperature thus removing the impurities (Gopinath et al., 2020). Total glycolysis refers to fully transforming the polyester into bis(2-hydroxyethyl) terephthalate ($C_6H_4(CO_2CH_2CH_2OH)_2$) (Tulashie et al., 2019). The purification then occurs through distillation. Total glycolysis produces polyester intermediates good enough to be used in industrial manufacture. The chain reaction is shown below: $(CO)C_6H_4(CO_2CH_2CH_2O)]_n$ + n $HOCH_2CH_2OH$ → n $C_6H_4(CO_2CH_2CH_2OH)_2$. Conversion through methanolysis produces dimethyl terephthalate (DMT). The chemical reaction is $[(CO)C_6H_4(CO_2CH_2CH_2O)]_n$ + 2n CH_3OH → n $C_6H_4(CO_2CH_3)_2$. Therefore, chemical recycling is aimed at degrading the polymers to make transformation possible.

8.7 APPLICATIONS OF RECYCLED PET

They are often titled rPET. The re-use can be similar to the original use. For example, a PET bottle made to contain water can continue to do so even after recycling. They can also be used as fibers. Most of them are sold as apparel fibers, including the fibers used to make carpets. Some companies such as Mohawk Industries have produced 100% PET-recycled fiber. Percentages can differ but PET can be used as fiber for textiles. Additionally, it can be used for energy recovery (Tulashie et al., 2019). Since PET contains hydrogen, carbon and oxygen, it can well be used as a source of fuel in waste-to-fuel plants.

8.8 POSSIBLE WAYS TO IMPROVE POTENTIAL HYGIENE PROBLEM BEFORE RECYCLING PET MATERIALS

Hygiene problem is solved through rigorous cleaning. They are separated from the metals. This is done to shield the granulator from damage. The metals can be present as a result of mixing the waste with other materials or they might have picked dirt during collection (Myren et al., 2020). It is good to make sure that the recycled materials do not have metals. The recycled materials are washed in hot water. Since these materials have been used, there is no doubt that they are dirty. Washing helps to make them clean. Washing is a simple mechanical process to clean the PET waste so that they can be processed. To get the best flakes from granules, flotation should be attempted, as it separates impurities from purities based on density. They can also be dried because drying removes impurities.

A simple cleaning process begins with opening the bale to release the contents. The contents are sorted because of the presence of contaminants of different colors or other polymers. For example, a bale can contain PET bottles of different colors, or different polymers such as PVC. Sorting should remove any foreign material including paper, film, sand, glass and stones, among others (Zhao et al., 2018). After sorting out the bale, the contents are rewashed. There are two ways to wash. One can cut the contents first or one can wash without cutting. After washing and drying, it is easy to see the remaining impurities with some being removed by air sifting. They can also be removed by placing the recycle materials in a container that will differentiate them using densities. Caustic washing can also help. Some of the cleaning processes are aimed at maintaining the intrinsic viscosity even if they decontaminate the materials to be recycled. Any process of washing should be followed by rinsing then drying. The washing makes things clean by removing dirt, but it cannot remove certain impurities, which is where processes such as air-sifting of flakes help.

There are other advanced steps that can help to further remove dirt. One of them is melt filtration, which is applied to remove contaminants from polymers melts during extrusion. The use and eventual recycling processes can lead to impurities and material defects. They accumulate with the length of the process. For example, reactive polyester OH$^-$ or COOH$^-$ end groups can die or become non-reactive (Shojaei et al., 2020). For example, this dying or inactivity of the end products can be seen when making vinyl ester end groups by dehydrating using terephthalate acid in which the OH$^-$ or COOH$^-$ end groups are likely to react with mono-functional degradation products like mono-carbonic acids or alcohols. This leads to minimized or no reactivity when the materials are being re-polycondensed. Sometimes the impurities are brought by the number of gels accumulating or long-chain branching defects. These deformities can be corrected through melt filtration. The chemical defect removal processes are meant to improve the quality of what is being recycled. Therefore, the manufacturer adds different elements for improved results. They can increase COOH$^-$ end groups, color, haze. Increment of transparent products reflects the original color of the PET polymers. They can even decrease mechanical properties such as the strength of the polymers or viscosity as necessary.

8.9 CHEMICAL STRUCTURES AND PROPERTIES OF DIFFERENT PET POLYMERS

Polyethylene terephthalate has two main polymers: the PET/PETE and PETG. The chemical formula for PET/PETE is $(C_{10}H_8O_4)_n$ or OOC-C$_6$H$_5$-COOCH$_2$-CH$_2)_n$. The molar mass varies in the range of 10–50 kg/mol and density is 1.38 g/cm^3, 20 °C 1.370 g/cm^3, amorphous 1.455 g/cm^3, single crystal (Nisticò, 2020). The melting point is > 250 °C (482 °F; 523 K). Therefore, PET is made up of polymerized units of the monomer ethylene terephthalate, with repeating $(C_{10}H_8O_4)$ units (Ng et al., 2018). It contains terephthalic acid. PET can be defined as a thermoplastic polymer resin of the polyester family, biodegradable and semi-crystalline (Nisticò, 2020). Crystalline polymers can be difficult to plasticize. Manufacturers must add water the same way as they do for nylon but to a lesser extent. The chemical formula shows that the polyester is based on carbon-carbon links in which one of the carbons forms part of a carbonyl group. PET is transparent to visible light and microwaves, and is quite effective in resisting ageing, wear and tear, and heat. It has less weight. It is least affected by impact and is shatter resistant. Furthermore, it emits good gas and possesses moisture barrier properties.

Being PET's copolymer, PETG contains both the PET and glycol properties. Consequently, it does not get as overheated as the PET (Ng et al., 2018). It is hard, chemical and impact resistant, transparent and ductile. Additionally, PETG can easily be extruded because it has good thermal stability, which makes it especially good for 3D printing. The extrusion temperature ranges from 220 °C to 260 °C with a print speed of 40–60 mm/s.

Since ABS 3D printing can cause PETG warping, it needs a heating plate, and is more prone to scratches than PLA. The heating plate ought not to reach 80 °C. PETG can attract moisture quickly and should be kept in a cool dry place.

Particular properties make PET/PETG ideal for 3D printing. They are strong and effective. They can withstand impact when being glazed or put on high-strength display units. It can easily be formed and can be injection molded into different shapes or extruded into sheets. It does not crack under pressure. Since it is naturally transparent, PETG can effectively accommodate other colors for special effects. However, it can be prone to oozing during 3D printing. Manufacturers will have to introduce bridging and retraction to remove blemishes. They can also do additional processing.

8.10 FEASIBILITY OF RECYCLED PET WASTE TO MAKE PET-BASED 3D PRINTING FILAMENTS

The recycled PET waste can make good PET-based 3D printing filaments. Extrusion method refers to a high volume manufacturing method whereby the plastic granulates are pushed down a cylinder with a rotating screw. Some printers have single screws while others have multiple screws. There are three zones in the cylinder, including the feeding zone, the melting zone and the pumping of the melted content zone. The feeding zone has a hoper which helps granulates or flukes to be fed into the cylinder (Woern et al., 2018). The feeding process is simple as the hopper is able to drop the granules using gravity. The only thing to ensure is that the drop is continuous because the end game is to have continuous strands of the recycled PET. This process should be started when cleaning, purification and decontamination have been done. As the granules drop, they move into the cylinder, a continuous hollow tube filled with screws. As soon as the granules drop they begin to melt. This cylinder normally has an external heating source. It is noteworthy that the heating should be gradual. To make plastic fibers, the solid flukes or granules have to be converted to a continuous pulp or semi-liquid material. This is why it is important to heat the granules to transform the shape into a continuous product. Excessive heating can melt the plastic too fast or too much to create amorphous hydrocarbons. The melting zone transforms into the maintaining zone where the melted granules must be kept in a certain shape. They are then pressed into a smaller opening to give the required shape. In this case, the formation of the filaments is the required shape.

The feasibility of the PET comes from the fact that it is strong enough to fit all the extrusion types. With right additives, one is sure to get the best filaments with PET polymers. For example, tubing extrusion is used to produce pipes and tubes. The filaments need to be long and strong. They come out of the screws long and strong. The strength of the PET enables the granules to stick together, probably with a little control of the intrinsic viscosity (Woern et al., 2018). The difference is that in tubing extrusion the continuous strands are only cooled after leaving the die, and normally, manufacturers use water cooling. Film extrusion is aimed at making sheets. The film tubes melt is cooled before it leaves the die and the semi-solid material formed is blown to expand to the right size and thickness. The suitability of the PET comes from its strength that prevents it from breaking. It is quite strong especially the co-polymer PETG, which is preferred for making film sheets. PET can also go through sheet film extrusion whereby the sheets are expanded after they leave the die by pulling because they are too thick to be blown away. Therefore, different rolls are used to pull the material apart to form the required film sheets. Even for those who want to do over jacket extrusion can use PET because it is effective in applying adhesive forces.

8.11 CONVERTING PET PLASTIC BOTTLES INTO FILAMENTS WHICH CAN BE 3D PRINTED

With tensile strength (σt) of 55–75 MPa, elastic limit of 50%–150%, notch test of 3.6 kJ/m^2, and glass transition temperature (Tg) of 67–81 °C, the PET polymers can withstand their conversion into filaments. The conversion process starts with setting up stepper motor connections. They pull the spools and provide a platform to roll them up. The process starts after the PET bottles have undergone certain transformations probably they have been cut and stranded together. The transparent strands are then passed through a nozzle of a small diameter. The nozzle compresses the plastic into a spool. The spools turn into thin filaments, which can then be printed. The 3D printing enables the making of products from PET without the requirement to mold. For example, in the video, a phone stand has been printed from the filaments. It is noteworthy that the PET has good adhesion and cohesion, and it spreads or sticks where needed thus enabling the formation of any shape.

8.12 MECHANICAL PROPERTIES CHANGE AFTER RECYCLING

Mechanical properties change when PET material is taken through recycling processes. The elongation viscosity increases. Usually, a chain extender is used to elongate the strands. The process produces a gel, and the more the chain extends, the more gel is formed and stickier the polymers become. The longer the chain the more immobile it becomes which shows how strong the influence of viscosity is. The elongation strength increases up to 0.6 wt% before it begins to decrease at the chain extender loading of 0.9 wt% (Alvarado Chacon et al., 2020). During recycling, the polymer can be extended but is not as strong as the virgin polymer even though the situation can be changed through the use of additives.

The tensile strength can also be affected. The virgin polymer of PET accommodates up to 11500 Pa of tensile strength while its copolymer PETG boasts of up to 7700 Pa (Esfandabad et al., 2020). This covers the stress or strain that a polymer can withstand before it breaks. It is expected to change because the recycled material cannot have a better tensile strength than the original one. Another change in mechanical property can be seen in the impact strength. This strength is measured by testing whether or not a material can break upon impact. PET's impact strength changes after recycling because the virgin polymers are stronger than the recycled ones. The changes will also cover the compression strength, which measures how long a product can take to withstand when pressed between two surfaces. PET has very impressive compression strength. It can be enhanced during recycling by adding carbon fibers or glass fiber reinforcements. Additionally, flexural strength can also be affected by recycling. The flexure test is conducted by bending a portion of PET material until a fracture shows. Since PET is a ductile material, it can withstand very high levels of flexural strength.

8.13 CONCLUSION

The global consumption rate of PET materials is stable and it is difficult to recycle the materials. The current recycling method and the chemical structure of PET materials were studied. The waste PET material is broken down to small granules and extruded to 3D printing filaments. Based on the test result, mechanical properties can achieve a reasonable and satisfactory result. Through the recycling of the PET materials to replace the PLA, the sustainability concept can be taught to individuals. Since PLA is obtained from sugar, potatoes or corns, using PET can be an alternative to save food.

ACKNOWLEDGMENT

The work described in this chapter was substantially supported by a grant from the Research Grants Council of the Hong Kong Special Administrative Region, China (UGC/FDS16/E05/19).

REFERENCES

Alvarado Chacon, F., Brouwer, M. T., & Thoden van Velzen, E. U. (2020). Effect of recycled content and rPET quality on the properties of PET bottles, part I: Optical and mechanical properties. *Packaging Technology and Science*, 33(9), 347–357.

Bainbridge, C. W. A., Engel, K. E., & Jin, J. (2020). 3D printing and growth induced bending based on PET-RAFT polymerization. *Polymer Chemistry*, 11(25), 4084–4093.

BNP Media. (2012, August 1). Market Trends: Developing Countries Spur Growth in Packaging Market. Retrieved April 3, 2022, from adgesivesmag: https://www.adhesivesmag.com/articles/91215-market-trends-developing-countries-spur-growth-in-packaging-marke

Esfandabad, A. S., Motevalizadeh, S. M., Sedghi, R., Ayar, P., & Asgharzadeh, S. M. (2020). Fracture and mechanical properties of asphalt mixtures containing granular polyethylene terephthalate (PET). *Construction and Building Materials*, 259, 120410.

Gopinath, K. P., Nagarajan, V. M., Krishnan, A., & Malolan, R. (2020). A critical review on the influence of energy, environmental and economic factors on various processes used to handle and recycle plastic wastes: Development of a comprehensive index. *Journal of Cleaner Production*, 274, 123031.

Khoo, K. S., Ho, L. Y., Lim, H. R., Leong, H. Y., & Chew, K. W. (2021). Plastic waste associated with the COVID-19 pandemic: crisis or opportunity?. *Journal of hazardous materials*, 417, 126108.

Myren, T. H., Stinson, T. A., Mast, Z. J., Huntzinger, C. G., & Luca, O. R. (2020). Chemical and electrochemical recycling of end-use poly (ethylene terephthalate)(PET) plastics in batch, microwave and electrochemical reactors. *Molecules*, 25(12), 2742.

Ng, G., Yeow, J., Chapman, R., Isahak, N., Wolvetang, E., Cooper-White, J. J., & Boyer, C. (2018). Pushing the limits of high throughput PET-RAFT polymerization. *Macromolecules*, 51(19), 7600–7607.

Nisticò, R. (2020). Polyethylene terephthalate (PET) in the packaging industry. *Polymer Testing*, 90, 106707.

Shojaei, B., Abtahi, M., & Najafi, M. (2020). Chemical recycling of PET: A stepping-stone toward sustainability. *Polymers for Advanced Technologies*, 31(12), 2912–2938.

Tulashie, S. K., Boadu, E. K., & Dapaah, S. (2019). Plastic waste to fuel via pyrolysis: A key way to solving the severe plastic waste problem in Ghana. *Thermal Science and Engineering Progress*, 11, 417–424.

Woern, A. L., Byard, D. J., Oakley, R. B., Fiedler, M. J., Snabes, S. L., & Pearce, J. M. (2018). Fused particle fabrication 3-D printing: Recycled materials' optimization and mechanical properties. *Materials*, 11(8), 1413.

Zander, N. E., Gillan, M., Burckhard, Z., & Gardea, F. (2019). Recycled polypropylene blends as novel 3D printing materials. *Additive Manufacturing*, 25, 122–130.

Zhao, Y. B., Lv, X. D., & Ni, H. G. (2018). Solvent-based separation and recycling of waste plastics: A review. *Chemosphere*, 209, 707–720.

Chapter 9

Additive Manufacturing Requirements for Dental Implementation

Upender Punia, Ashish Kaushik, Ramesh Kumar Garg, Anmol Sharma, and Deepak Chhabra

CONTENTS

9.1 INTRODUCTION

Additive manufacturing (AM) (colloquially 3D printing) is characterized as a process of fabricating parts from three-dimensional (3D) model data (CAD file), generally by incremental layer deposition, as opposed to conventional

DOI: 10.1201/9781003306238-9

manufacturing techniques (Sharma et al., 2022). Three-dimensional printing is the subject of long-standing interest to researchers and dentists including virtual surgical planning. The technological advancements are seemingly on the near horizon. The digital transformation of dentistry, including milling of ceramic and composite materials, intra and extra-oral scanning, and computerized technology, is firmly established. Rapid prototyping technologies to develop digitized models of human anatomical parts are creating a shift in how products are designed and fabricated (Telfer and Woodburn, 2010). Modern-day 3D printers can control the massive demand for temporary or provisional restorations and equipment for achieving the expertise required by dental professionals. Additive manufacturing processes like stereolithography, powder binding, photopolymer jetting, digital light processing, fused deposition modelling (FDM), direct metal laser sintering, etc., are future-proof technology that significantly improves dentistry workflow. It can drastically decrease patient chair time and offer flexibility in customized products, better quality, and accuracy in 3D printed parts. The most recent and wide range of 3D printing applications include fixed or removable prosthodontics, orthodontics, endodontics, maxillofacial surgery, implant dentistry, and periodontics. The chapter gives an overview of various additive manufacturing technologies used in dental practices and their applications in dentistry.

9.2 NEED FOR DIGITAL MANUFACTURING AND 3D PRINTING IN DENTISTRY

Dental technology, like other technologies, is also influenced by the fundamental changes occurring in society, such as scarcity of skilled labour and a continuous decrease in the number of trainees in dental practices. However, the need for dental prostheses is high due to varying demographics (Schweiger et al., 2021; Campbell et al., 2017). Conventional dentistry techniques significantly impacted the costs, quality, efficiency, and ability to interact with dental technicians and surgeons. Digital processes are preferred due to their efficiency, accuracy, reproducibility, and enhanced material properties. Dental professionals firmly believe that digital technology continues to advance the field of dentistry, and going digital is the way forward. Digital manufacturing proves to be an asset for dental professionals as it allows them to stay updated with the recent technology and dental materials. Mass production, part customization, and complex design fabrication are more accessible in 3D printing than traditional subtractive techniques. A digitally fabricated product is more readily available, economical, and paves the path for digital dentistry (Van Noort, 2012; Horn and Harrysson, 2012; Caviezel et al., 2017; Kessler et al., 2020a).

9.3 CURRENT STATE OF DENTAL 3D PRINTING

The rapid prototyping technique has been firmly accepted in dental practices for nearly two decades, for instance, the laser sintering technique of Bego Medical (Bremen, Germany) and EOS (Krailing, Germany). During its infancy stage in 2002, the process of printing metals induced a sensation. Experts acknowledged the vast potential of this rapid prototyping technology. Seletive laser melting (SLM) is widely used to fabricate metallic structures, as stereolithography is also used in dental applications to manufacturing surgical templates subjected to point-by-point solidification within a resin vat through a laser source. In the early years, 3D printers used for dental practices were confined to large manufacturing centres and industries. In today's scenario, these printers can be accessed by "regular" dental laboratories and empower dental experts and help them tackle daily challenges. Also, the number of vendors entering the dental market and providing rapid prototyping techniques has drastically increased. Dental professionals can now quickly fabricate dental models and other occlusal splints, which was once unimaginable. Dental laboratories can quickly fabricate objects using acrylics or composite resin in preparatory stages like dentures, crowns, etc. Rapid prototyping of patient specific organs or tissue is a becoming a viable alternative owing to the advancement in biomaterials and 3D printing techniques and these processes develops enormous opportunities to change the existing structure of manufacturing system (Vashistha et al., 2019; Punia et al., 2022).

9.4 TECHNIQUES OF ADDITIVE MANUFACTURING PROCESS IN DENTISTRY

The different additive manufacturing processes are used in dentistry concerning their applications. Every fabrication technique has its specific quality, i.e., stereolithography (SLA) is famous due to its accuracy and precession; digital light processing is very fast in the fabrication of products; direct energy deposition can print the metal part resulting in the fabrication of strengthened parts, and fused deposition modelling printers are cheaper as compared to other printers. The different types of rapid prototyping techniques are explained in this section of the chapter, including their advantages, limitations, and applications in dentistry. Figure 9.1 demonstrates the various fundamental additive manufacturing techniques used in dentistry.

9.4.1 Stereolithography

From a technological perspective, stereolithography (SLA) can also be called photopolymerization with thermoset plastics. It contains a thermosetting polymer or resin that is hardened irreversibly by curing a soft solid

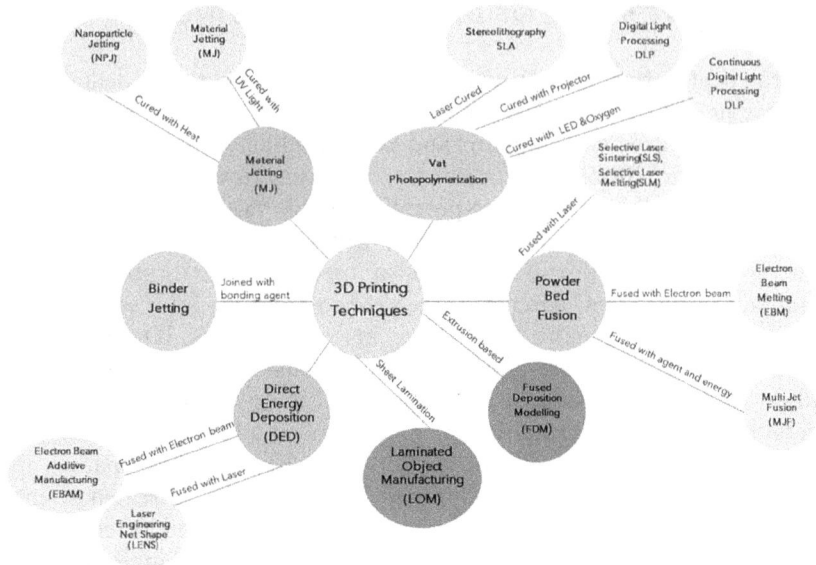

Figure 9.1 Various AM techniques used in dentistry.

or viscous liquid polymer or resin. Heat or sufficient radiation causes curing, which can be accelerated by applying high pressure or combining with a catalyst. The origin of the stereolithography apparatus can be traced back to the 1980s, the first patented and commercially available 3D printer for rapid prototyping (Zaharia et al., 2017).

In simple terms, stereolithography works because a photosensitive resin gets solidified when exposed to ultraviolet (UV) radiations (Negruţiu, 2017). A scanning laser is used in stereolithography, and UV light is aimed to a photosensitive resin surface (Anderson et al., 2018a). The layers bind and fabricate the prototype, starting at the bottom and building upwards. This technique can fabricate surgical guides, obturators, crowns, and partial or removable dentures. The limitation of this process lies in the inadequacy of biocompatible resin and increased cost during the fabrication of large size parts (Jain et al., 2016; Chia and Wu, 2015). Other challenges of SLA include the utilization of photoinitiators and radicals, which may concede the post-processing time, cytotoxicity, and inducing entrapment of un-reacted monomers within itself (Dawood and Marti, 2015). The schematic working of the SLA process is demonstrated in Figure 9.2.

9.4.2 Digital light processing

Digital light processing (DLP) is another technique of additive manufacturing. Its printing methodology is very similar to that of Stereolithography. Larry Hornbeck of Texas Instruments (TI) in the year 1987 created this technique.

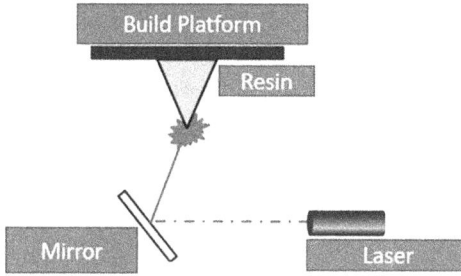

Figure 9.2 Working of stereolithography 3D printing.

This methodology became very famous as a projector is used to produce models. A liquid polymer container is used in DLP. The light from a projector in the setup falls on this vat. This process is carried out under feasible light conditions. As demonstrated below, the projector projects a virtual picture of the 3D structure or model onto the liquid polymer. The liquid polymer hardens when exposed to light, and then the construction platform slides downward, exposing the liquid polymer to light once more. This technique is repeated until the entire geometry of the 3D construction has been built. Finally, the liquid is withdrawn from the container, revealing the solidified model. Principle motion is shown in Figure 9.3. The DLP method is fast, and it can print high-resolution objects (Raza and Singh, 2020).

9.4.3 Selective laser sintering

The University of Texas invented selective laser sintering (SLS) and DTM Corporation was the first firm to commercialize it (laterally owned by 3D

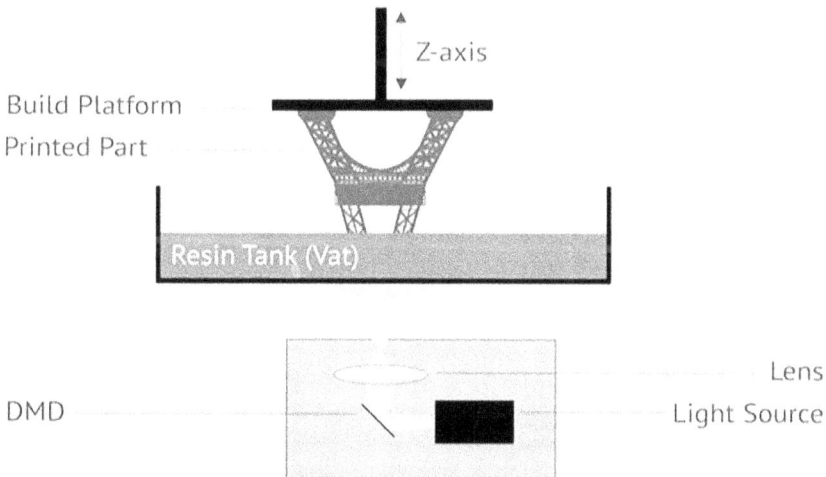

Figure 9.3 Digital Light Processing Setup.

systems) (Sandeep and Chhabra, 2017). In SLS, the three-dimensional product is the sintered part, i.e., the fused part. In Selective Laser Sintering (SLS), a 3D effect is formed using laser-sinterable powder. The material that has not been sintered or that has not been run through a laser to harden in each layer may even serve as a support structure for the actual component. This may be swept off after the manufacture is finished (Sood, 2011). Here the powder is introduced into the laser sintering machine, and the products formed by this method have very high tensile strength, and the products possess very efficient thermoplastic properties. Laser of high power is used in selective laser sintering technique to fuse small particles of material used for manufacture, plastic or metal. This laser firstly scans the area of cross-sections generated using computer-aided drawing (CAD) onto the power bed and accordingly fuses the powdered material. Laser power, temperature, and part orientation are vital in determining fabricated parts' strength and surface quality (Kumar et al., 2016).

After fusing, as done in previous methods, the fused layer gets lowered down to its entire thickness. Subsequently, the following layer formation material is applied, and it comes over it. This layer deposition method lowers the power bed and then another layer deposition until the complete product is formed. Figure 9.4 illustrates the working of the SLS process.

If the materials are used for the SLS method, it is to note that high-strength composites are not suitable because microscale reinforcements (millimetre length fibres) tend to arrive at high strength. The powder layer manufacturing method is limited to particle size, nearly 100-micrometer diameter, or even less.

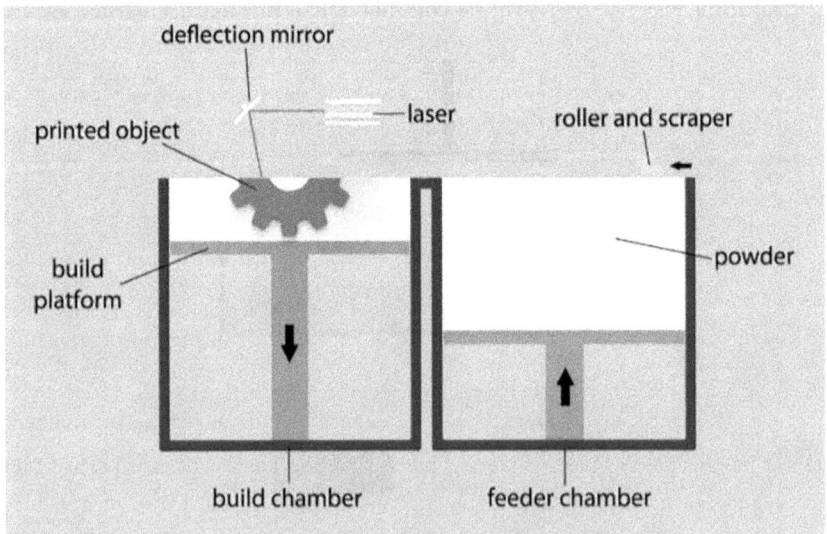

Figure 9.4 Schematic view of the SLS technology.

9.4.4 Selective laser melting

This method of manufacturing can also be stated as Directed-Energy-Deposition. It can generate three-dimensional objects using a laser beam supplied to metal powder. The material used is metal powder, which has been directed by the high-power beam of laser light for its deposition leading to the formation of molten build material of the required design. As this process involves the use of laser for manufacturing, this method of 3D printing is also called by one other name as 'Laser Engineered Net shaping'. This method involves using an elevated power-density laser beam for melting and later fusing the metallic powders. This manufacturing technique somehow acts like SLS and, i.e., why it is sometimes called its subcategory. This method shows that additive manufacturing has the potential to melt the metallic powder completely and then fuse it in the shape of desired solid 3D part (Cabrera Agudo, 2018). Figure 9.5 shows a schematic depiction of the SLM process. However, there are various similarities between SLS and SLM; there are differences too. One main difference is that SLS uses different materials like glass, plastic, and ceramics while SLM only uses metallic material, which means that SLM powder is not being fused; instead, it is firstly liquefied for enough time also to turn metal into a homogeneous molten state. Some materials used for manufacturing in SLM are copper, tool steel, titanium, tungsten, Al, (gold) Au, stainless steel, etc. A suitable property of this is its reduced porosity, which controls its crystal structures. SLM can build parts

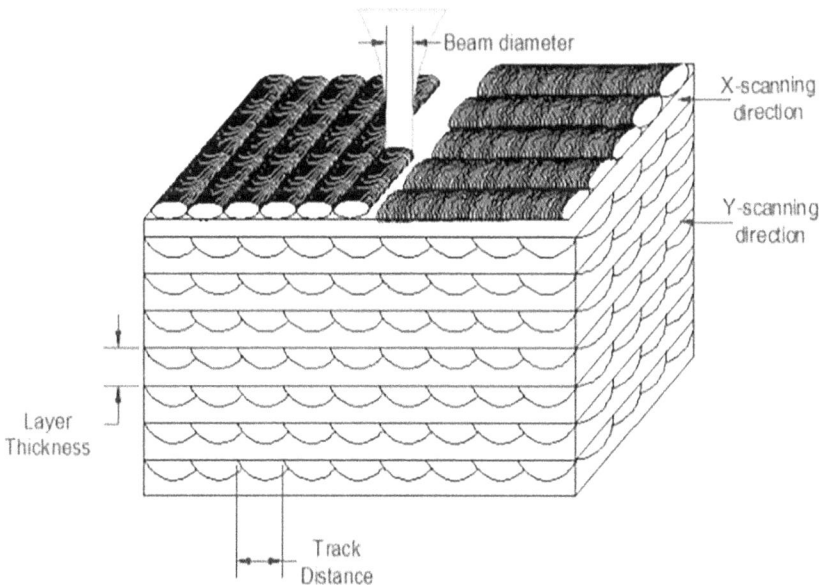

Figure 9.5 Schematic representation of SLM process.

stronger than SLS. This prevents part failures. The only limitation in SLM is using a single metal powder (Paudyal, 2015).

9.4.5 Fused deposition modelling

FDM is the process developed and patented by Stratasys Inc. Parts are fabricated. It consists of an extruder that extrudes the material on a surface, and the surface moves in three linear axis motions according to the desired pattern. This movement is usually driven by stepper or servo motors. The material is continuously fed into the printing head, which moves in cartesian coordinates (X & Y coordinates) to print the initial layer of parts designed in CAD software by the depositing material. The printer head base is taken down to fabricate more layers. Fabricated parts are either subjected to a support removal solution to remove the supporting material or can also be performed manually. Thermoplastics like PLA, ASA, Onyx, etc., are extensively used for the FDM process due to their low melting temperature (Wang et al., 2017). Parts fabricated by this technique generally possess better fracture toughness and tensile strength. However, the strength of parts fabricated by this technique is significantly affected by build orientation, part geometry (Gardan et al., 2016; Galeta et al., 2016). Figure 9.6 illustrates the extrusion based FDM process.

9.4.6 Electron beam melting

In this rapid prototyping technique, heat generated from the electron beam has been utilized to solidify the selective areas of the powder bed where it is desired to create 3D parts. This process fabricates parts using an electron

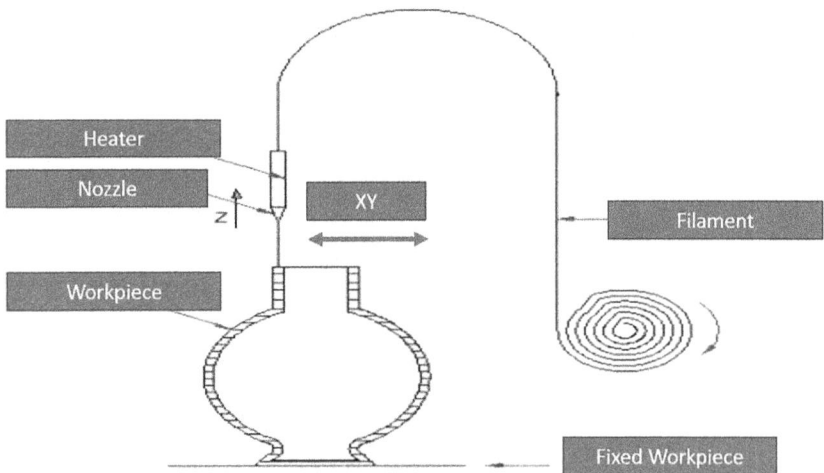

Figure 9.6 Extrusion based FDM Technology.

beam instead of a laser source (Sawhney and Jose, 2018). Here electron beam is utilized to build 3D elements in a vacuum. This is used for metal parts. The raw material may be metal powder or wire. The construction material is initially pre-heated to an ideal temperature utilizing the first pass of the electron beam in printing 3D components. The object layer's pattern is melted with the second pass of the electrical beam. Now, all electron beam passes are used to soften the material present in bulk inside the outline (Raza and Singh, 2020). Figure 9.7 depicts the primary procedure.

This method of manufactured parts has high density and good mechanical properties. The materials used are titanium (Ti) and some of its alloys, steel, nickel alloys, etc. This rapid prototyping technique has found enormous applications in oral, maxillofacial surgery, and orthopedics, where attention is focused on the fabrication of customized implants in porous scaffolds form (Hung et al., 2016).

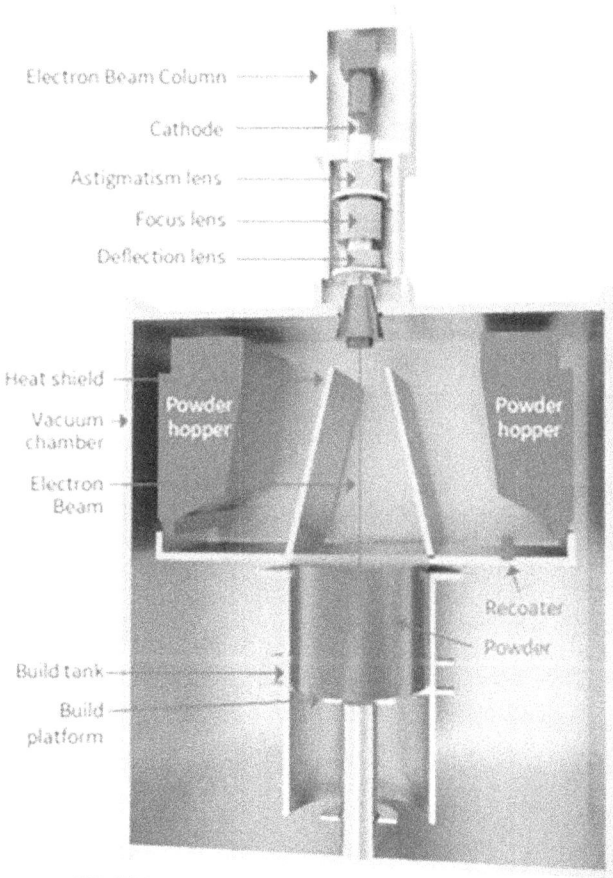

FIGURE 1

Figure 9.7 Electron beam melting process.

9.4.7 Laminated object manufacturing

This process was invented in 1991 by a corporation named 'Helisys' in California, USA. Its principle is distinct from other techniques of 3D printing because of its working principle of 'Sheet Lamination.' In this manufacturing method, objects are built layer over layer by sticking laser sheets of material together; a schematic for this technique is shown in Figure 9.8. These materials can be paper, metal foil, or plastic. Unlike previous printing techniques, it doesn't work on a powder-based mechanism, and therefore it doesn't require any support structure (Gurrala and Regalla, 2017).

9.4.8 3D Inkjet bioprinting

Direct inkjet printing (DIP) technology originated at the Massachusetts Institute of Technology (MIT), gaining popularity due to its fabricating of dense structures with intricate designs. The key distinction between inkjet printing and other rapid prototyping methods is that inkjet printing can make thick structures, while all other rapid prototyping processes can only produce porous structures. To accomplish solidification, a coating of powder is evenly applied to the storage and mixed with drops of binding material printed into the surface. In this work, Ebert et al. use the DLP approach to create zirconia prostheses for dental implants (Sheela et al., 2021). The process is generally more proficient with the most minor material consumption. Heat or transient pressure is used to generate ink droplets, allowing ink to flow out of the nozzle and onto the surfaces. Heat causes non-uniform and insufficiently mixed droplets, resulting in rough surfaces, whereas directional printing with

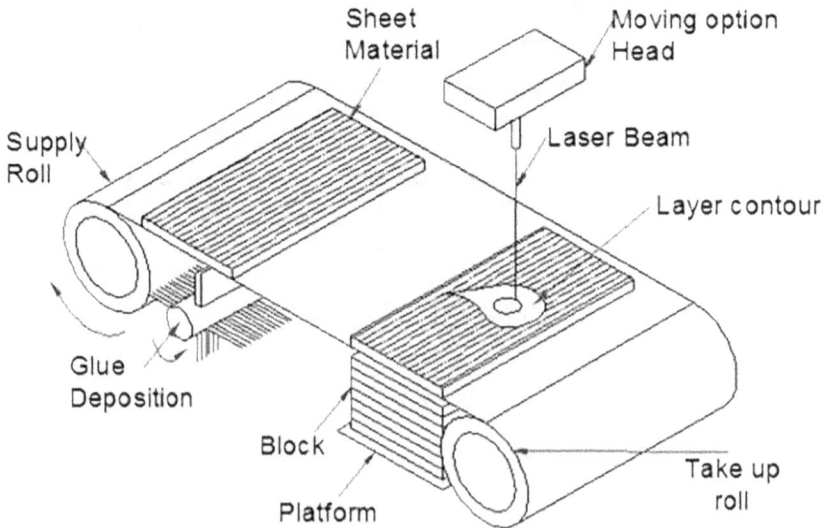

Figure 9.8 Illustration of the laminated object manufacturing process.

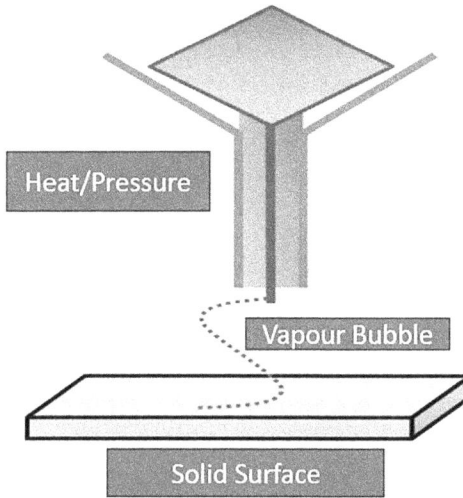

Figure 9.9 3D Inkjet printing.

the uniform size is achieved by the latter. The method has found widespread use in the fabrication of porous calcium phosphate (CPP) frameworks for biomedical applications. The University of Sheffield implements the technique for producing colored soft-tissue prostheses. Although, this rapid prototyping technique is not used in dental practices due to certain limitations. One of this technique's drawbacks lies in the toxic nature of polymer glues, which restricts the parts fabricated by this technique for use in biological applications. The heating/pressure 3D inkjet is depicted in Figure 9.9.

9.5 REQUIREMENTS AND APPLICATIONS OF 3D PRINTING IN DENTISTRY

3D printing has arrived in dentistry, intending to bring aesthetic dentistry accessible to the hands of all practitioners and provide comfort, function, and speech to the patients (Jain et al., 2018). The technology aims to bring restorative processes functional and artistic control into the chairside setting. Almost every aspect of dentistry, such as dental prosthesis, surgical templates, occlusal splints, aligners, metal frameworks, provisional or permanent restorations, and removable dentures, are covered by 3D printing. Vital applications of AM in dentistry are illustrated in Figure 9.10.

9.5.1 3D printing in prosthodontics

Three-dimensional printing plays a vital role in both fixed and removable prosthodontic applications. Provisional and permanent indirect restorations,

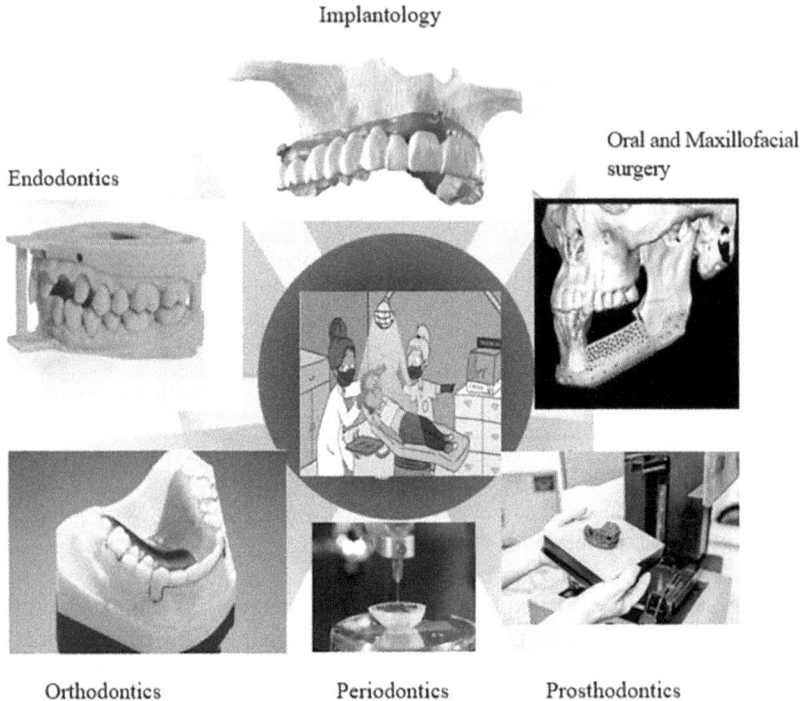

Figure 9.10 3D printing applications in dentistry.

including crowns, bridges, and permanent monobloc direct restorations, can be easily customized and fabricated. In the case of removable prosthodontics like, digital occlusal designs are delivered quickly. Customized trays can be manufactured by computerized scan models or impressions (Zaharia et al., 2017). Provisional crowns and bridges fabricated with resin 3D printing exhibit considerably greater accuracy enhanced mechanical properties and marginal fit as compared to traditional techniques (Ishida and Miyasaka, 2016).

9.5.2 Crowns and bridges

In prosthodontics, lost wax crowns and bridges are often employed. Conventional methods used to manufacture crowns and bridges are comparatively more inclined to human error and require extensive labor. In his study, Mai et al. (2017) found that additive manufacturing and milling processes show better and more accurate marginal fits than manual methods. Three-dimensional printed crowns have a nearly perfect occlusal fit and the fewest possible discrepancies (Mai et al., 2017). When all entire line designs were manufactured via 3D printing, there were less marginal gaps (Alharbi et al., 2018). Several other studies also concluded that crowns fabricated by 3D printing exhibit significant marginal fit (Yildirim, 2020). Prechtel et al. (2020) evaluated the mechanical qualities of items produced utilizing different prototype approaches in their study (Papadiochou and Pissiotis, 2018). The research looked at the fracture stress of 3D printed components created using PEEK material versus traditional techniques, composite restorations, and a variety of other parameters. The study's findings demonstrated that 3D printed patterns stayed unchanged even after fracture force testing, resulting in higher success rates than conventional direct restorations (Prechtel et al., 2020). A wide range of materials with various mechanical properties can be used for crown fabrication, including ceramics (alumina, zirconia), metals, polymers, etc. are used due to their greater bond strength, invariable mechanical properties process used for fabricating zirconia implants demonstrates considerable dimensional accuracy (Osman et al., 2017). According to an in-vitro research, 3D printed zirconia crowns have the same trueness as automated CAD/CAM crowns, demonstrating zirconia's value as a dental material for prosthodontic applications (Wang et al., 2019).

9.5.3 Denture

There is a substantial growth in implementing digital manufacturing in prosthodontic applications, such as the fabrication of partial and complete dentures (McLaughlin et al., 2019). Due to a diverse range of CAD/CAM processes, the method for manufacturing dentures can vary accordingly and leads to considerably reduced chair time for patients (Saponaro et al., 2016). Digital dentistry is capable of electronic data storage that empowers dental technicians to fabricate multiple dentures in short time duration and superior quality (Kalberer et al., 2019). The digital workflow method starts with digital impressions, which may be done immediately with an intraoral scanner or with extra-oral scanning of models created from the impressions (Papaspyridakos et al., 2016). The typical approach for taking impressions involves the use of alginate impression materials, which may cause choking, mucosal irritation, and discomfort for the patient. Digital impressions, on the other hand, need less time to complete and have less flaws (Wilk, 2015). For denture design, a digital file such as Standard Tessellation Language (STL) may be submitted to CAD software (Clark et al., 2019). However, limitations for intraoral scanning lies in obstacle in the acquisition of clear scan due to the presence of various fluids in the mouth (Wilk, 2015). The capacity of intraoral scanners to reproduce edentulous arches was examined in the research. The findings show that matching digital impressions are feasible; nevertheless, the scanners' accuracy varies greatly (Patzelt et al., 2013).

In a few cases, a well-designed removable partial denture (RPD) used to restore the lost tooth is the only solution, like in the occurrence of long edentulous spans, lost residual ridges, or unavailability of posterior abutments (Hu et al., 2019; Bajunaid et al., 2019). A 2017 article demonstrates how 3D printing may be used to create a pure titanium metal construction for a detachable partial denture of the maxillary arch (Hu et al., 2019). For constructing detachable partial dentures with a distal extension, Wu et al. (2020) discussed the utilization of several fast-prototyping techniques. The stereolithography technique is used to fabricate customized trays for the patients, while the SLM method is used for printing metal frameworks (Wu et al., 2020). Several in-vitro studies demonstrate SLM parts' identical fit accuracy compared to the lost wax technique (Hu et al., 2019; Negm et al., 2019). Moreover, parts manufactured by SLM represent a better fit than the conventional milling method (Presotto et al., 2019).

9.5.4 3D printing in endodontic

The pioneered 3D printing has revolutionized the design possibilities for constructing novel restorations, surgical templates, and dental models, wildly flourishing the success rate of traditional surgeries (Ahn et al., 2018). However, the scope of 3D printing in endodontics is not vastly explored, and various clinical studies determine the improvements brought by 3D printing in guided access, auto transplantation, endodontics, and standard dental educational practices (Anderson et al., 2018b).

One of the important uses of rapid prototyping in the restoration of dental and craniofacial tissues is the fabrication of complex 3D structures to replicate a functional extracellular matrix. The requirements for regenerative construction are available, but biomanufacturing techniques are confined (Giacomino et al., 2018). Pulp extracted bleeding is mixed with the body's clots for root canal treatment, which controls remodeling after a process. Although, promising results are obtained in techniques using bioinspiration methods such as novel bioinks (Athirasala et al., 2018). Several research has proposed that innovative bioinks might be used as a matrix structure that matches their natural constitution. For example, Athirasala et al. demonstrated the development of Alg-Dent, a new bioink, using printable alginate hydrogels, which are then blended with various dentin matrix concentrations. The study revealed that insoluble dentin matrix and generally high levels of alginate are ideal for cell viability.

Moreover, odontogenic differentiation potential is significantly improved with the high concentration of soluble dentin matrix, acting as an ideal biomaterial in regenerative dentistry (Athirasala et al., 2018). Three-dimensional printed resins are widely used in root canal filling treatments, including the filling of imitating C-shaped channels (Gok et al., 2017). The practical implementation of rapid prototyping in endodontics can ultimately improve the results of root canal treatments and endodontic microsurgeries in patients.

9.5.5 3D printing in oral and maxillofacial surgery

The application of 3D printing in Oral and Maxillofacial Surgery (OMFS) clinical applications has been embraced over the few years due to widespread technology and openness to economical 3D printers (Louvrier et al., 2017; Khorsandi et al., 2021). Oral surgeons use 3D printing and virtual planning to treat patients effectively. In their study, Jacobs and Lin (2017) classified the four effective uses of 3D printing during craniomaxillofacial operations as contour models (Type I), surgery guides (Type II), splints (Type III), and implants (Type IV) (Jacobs and Lin, 2017). Fabrication of contour models, also referred as positive space models, are primarily used as they directly print the object depending upon the external anatomy of the patient. Models can be easily fabricated using in-house, low-cost, economical 3D printers, resulting in timesaving in alarming situations like fractures (Lin and Yarholar, 2020). In cases where autologous bone grafts are the sole option for replacing damaged structures, the technology also helps with reconstruction plates and surgical guides (Largo and Garvey, 2018). Splints for orthognathic rectification, including jaw sequences and occlusions, are classified as Type III. Type IV categories consisting of implants are generally less developed than the 3D printed parts because of higher demands during the fabrication. Such implants are precise from a functional, structural, and biological point of view. They are utilized for jaw reconstruction and cranial repairs to provide necessary support and shape.

9.5.6 3D Printing in dental implants

Dental implantation is the preferred solution for patients suffering from tooth loss (Bollman et al., 2020). Dental implants are generally positioned in the jawbone with the help of a surgical procedure and will anchor themselves through osseointegration (Alghamdi, 2018). It is the method by which opposition of bone takes place among the interface of bone-implant (Altay et al., 2018). Metallic implants are generally used on a larger scale in dentistry as a substitute for lost teeth, giving rise to rapid prototyping.

Rapid prototyping allows for the manufacturing of accurate and economic implants used in dental applications (Dalal et al., 2020). The procedure starts with creating a 3D CAD model, which is then split into cross-sectional components and transmitted to a 3D printer, which fabricates the part in layers (Nesic et al., 2020). It helps manufacture dental implants with intricate geometry, surgical template, and drill guides. Rapid prototyping's capacity to print bone tissues allows it to be used as a biomimetic substrate inside the mouth to promote bone cell proliferation, differentiation, and growth (Heo et al., 2017). Moreover, priorly fabricated surgical guides for dental implants can help locate or verify the appropriate location (Gittelson, 2008). A brief overview of various materials used in dental implants is discussed in Table 9.1.

Table 9.1 Brief overview of 3D printed dental implant materials

S. No.	Materials	Literature studies
1	Plastic	(Kalman, 2018)
2	Zirconia	(Osman et al., 2017)
3	Titanium	Tedesco et al., 2017
4	Acrylate	(Bianchi et al., 2020; Mangano et al., 2020)
5	Poly-ether-ether-ketone	(Yang et al., 2019; Han et al., 2019)
6	PEEK blended Magnesium Phosphate	(Ferreira et al., 2020; Sikder et al., 2020)
7	Cobalt-Chromium Alloy	(Hong et al., 2020; Bae et al., 2020)

9.6 CURRENT ISSUES AND FUTURE PROSPECTS

Additive manufacturing techniques are gaining an edge in various aspects of dental practices and differ vastly from subtractive manufacturing techniques. The technology is comparatively more accurate, produces negligible waste, and can process different materials, including ceramics, metals, and plastics applicable in dentistry (Kessler et al., 2020b). Although 3D printing is fundamentally changing the dynamics of dentistry by promoting an entirely digitized clinical workflow, it still has a long way to go. For consideration, CT scans incorporated with 3D printing for surgical operations lead to the fabrication of surgical guides. Still, their use is restricted due to the inability of certain materials lacking sterilization (Dawood et al., 2015). Moreover, accuracy is significantly affected by the acquisition of data by the intra-oral scanners, which are generally less accurate during scanning irregular surfaces or full arch scans (Abduo and Elseyoufi, 2018). Intraoral scanner dental practices are continuously transitioning to digital manufacturing due to expanding demographics and the simple availability of 3d printers, resulting in effective patient diagnosis and treatment with reduced chair time. The transfer of technology also reduces storage requirements. Three-dimensional printing also gives rise to ethical issues, including data protection, privacy, and confidentiality, primarily because data digitization can make it easily accessible for research considerations (Favaretto et al., 2020). Moving forward, this transformative rapid prototyping process needs to conquer the dental industry's skepticism regarding novel printable materials such as permanent restorations and printed dentures.

9.7 CONCLUSION

When incorporated with computerized CAD/CAM and CBCT, rapid prototyping leads to a significant shift to digitized dentistry, which is revolutionizing the dental profession. A dentist's main task is moving to a digital workflow and consolidating these explosive techniques into regular practice.

It enables the dentist to stay updated with modern techniques resulting in cost-effective, less invasive patient treatment. Modern-day 3D printers are proficient in tackling the massive demand for dental restorations with negligible waste and help achieve adequate clinical excellence. Precise scanning of patients with intraoral scanning empowers the capability to print customized models with higher accuracy levels. Although there is a growing acceptance of this transformative technology, new standards using the equipment must be established, keeping in mind the patient's health, care, and safety. Rapid prototyping, a new technology, follows a learning curve; dental professionals are advised to begin with simple processes and tackle the complex procedure after gaining exposure and experience.

REFERENCES

Abduo, J., & Elseyoufi, M. (2018). Accuracy of intraoral scanners: a systematic review of influencing factors. *The European Journal of Prosthodontics and Restorative Dentistry*, 26(3), 101–121.

Ahn, S. Y., Kim, N. H., Kim, S., Karabucak, B., & Kim, E. (2018). Computer-aided design/computer-aided manufacturing–guided endodontic surgery: guided osteotomy and apex localization in a mandibular molar with a thick buccal bone plate. *Journal of Endodontics*, 44(4), 665–670.

Alghamdi, H. S. (2018). Methods to improve osseointegration of dental implants in low quality (type-IV) bone: an overview. *Journal of Functional Biomaterials*, 9(1), 7.

Alharbi, N., Alharbi, S., Cuijpers, V. M., Osman, R. B., & Wismeijer, D. (2018). Three-dimensional evaluation of marginal and internal fit of 3D-printed interim restorations fabricated on different finish line designs. *Journal of Prosthodontic Research*, 62(2), 218–226.

Altay, M. A., Sindel, A., Özalp, Ö., Yildirimyan, N., Kader, D., Bilge, U., & Baur, D. A. (2018). Does the intake of selective serotonin reuptake inhibitors negatively affect dental implant osseointegration? A retrospective study. *Journal of Oral Implantology*, 44(4), 260–265.

Anderson, J., Wealleans, J., & Ray, J. (2018a). Endodontic applications of 3D printing. *International Endodontic Journal*, 51(9), 1005–1018.

Anderson, J., Wealleans, J., & Ray, J. (2018b). Endodontic applications of 3D printing. *International Endodontic Journal*, 51(9), 1005–1018.

Athirasala, A., Tahayeri, A., Thrivikraman, G., Franca, C. M., Monteiro, N., Tran, V., & Ferracane, J. (2018). A dentin-derived hydrogel bioink for 3D bioprinting of cell laden scaffolds for regenerative dentistry. *Biofabrication*, 10, 024101. [CrossRef]

Bae, S., Hong, M. H., Lee, H., Lee, C. H., Hong, M., Lee, J., & Lee, D. H. (2020). Reliability of metal 3D printing with respect to the marginal fit of fixed dental prostheses: A systematic review and meta-analysis. *Materials*, 13(21), 4781.

Bajunaid, S. O., Altwaim, B., Alhassan, M., & Alammari, R. (2019). The fit accuracy of removable partial denture metal frameworks using conventional and 3D printed techniques: an in vitro study. *Journal of Contemporary Dental Practice*, 20(4), 476–481.

Bollman, M., Malbrue, R., Li, C., Yao, H., Guo, S., & Yao, S. (2020). Improvement of osseointegration by recruiting stem cells to titanium implants fabricated with 3D printing. *Annals of the New York Academy of Sciences, 1463*(1), 37–44.

Cabrera Agudo, B. (2018). *Analysis of Additive Manufacturing in the Aeronautical Field* (Master's thesis, UniversitatPolitècnica de Catalunya).

Campbell, S. D., Cooper, L., Craddock, H., Paul Hyde, T., Nattress, B., Pavitt, S. H., & Seymour, D. W. (2017). Removable partial dentures: the clinical need for innovation. *The Journal of Prosthetic Dentistry 118*(3), 273–280.

Caviezel, C., Grünwald, R., Ehrenberg-Silies, S., Kind, S., Jetzke, T., & Bovenschulte, M. (2017). Additive Fertigungsverfahren (3-D-Druck). Innovationsanalyse.

Chia, H. N., & Wu, B. M. (2015). Recent advances in 3D printing of biomaterials. *Journal of Biological Engineering, 9*(1), 1–14.

Clark, W. A., Duqum, I., & Kowalski, B. J. (2019). The digitally replicated denture technique: a case report. *Journal of Esthetic and Restorative Dentistry, 31*(1), 20–25.

Dalal, N., Ammoun, R., Abdulmajeed, A. A., Deeb, G. R., & Bencharit, S. (2020). Intaglio surface dimension and guide tube deviations of implant surgical guides influenced by printing layer thickness and angulation setting. *Journal of Prosthodontics, 29*(2), 161–165.

Dawood, A., & Marti, B. M. (2015). Sauret-Jackson V, Darwood A. 3D Printing in Dentistry Brit. *British Dental Journal, 219*(11), 521–529.

Dawood, A., Marti, B. M., Sauret-Jackson, V., & Darwood, A. (2015). 3D printing in dentistry. *British Dental Journal, 219*(11), 521–529.

Favaretto, M., Shaw, D., De Clercq, E., Joda, T., & Elger, B. S. (2020). Big data and digitalization in dentistry: a systematic review of the ethical issues. *International journal of Environmental Research and Public Health, 17*(7), 2495.

Galeta, T., Raos, P., Stojšić, J., & Pakši, I. (2016). Influence of structure on mechanical properties of 3D printed objects. *Procedia Engineering, 149*, 100–104.

Gardan, J., Makke, A., & Recho, N. (2016). A method to improve the fracture toughness using 3D printing by extrusion deposition. *Procedia Structural Integrity, 2*, 144–151.

Giacomino, C. M., Ray, J. J., & Wealleans, J. A. (2018). Targeted endodontic microsurgery: a novel approach to anatomically challenging scenarios using 3-dimensional–printed guides and trephine burs—a report of 3 cases. *Journal of Endodontics, 44*(4), 671–677.

Gittelson, G. (2008). *U.S. Patent Application No. 11/933,815.*

Gok, T., Capar, I. D., Akcay, I., & Keles, A. (2017). Evaluation of different techniques for filling simulated C-shaped canals of 3-dimensional printed resin teeth. *Journal of Endodontics, 43*(9), 1559–1564.

Gurrala, P. K., & Regalla, S. P. (2017). Friction and wear rate characteristics of parts manufactured by fused deposition modelling process. *International Journal of Rapid Manufacturing, 6*(4), 245–261.

Han, X., Yang, D., Yang, C., Spintzyk, S., Scheideler, L., Li, P., ... & Rupp, F. (2019). Carbon fiber reinforced PEEK composites based on 3D-printing technology for orthopedic and dental applications. *Journal of Clinical Medicine, 8*(2), 240.

Heo, E. Y., Ko, N. R., Bae, M. S., Lee, S. J., Choi, B. J., Kim, J. H., ... & Kwon, I. K. (2017). Novel 3D printed alginate–BFP1 hybrid scaffolds for enhanced bone regeneration. *Journal of Industrial and Engineering Chemistry, 45*, 61–67.

Horn, T. J., & Harrysson, O. L. (2012). Overview of current additive manufacturing technologies and selected applications. *Science Progress*, 95(3), 255–282.

Hu, F., Pei, Z., & Wen, Y. (2019). Using intraoral scanning technology for three-dimensional printing of kennedy class i removable partial denture metal framework: a clinical report. *Journal of Prosthodontics*, 28(2), e473–e476.

Hung, K. C., Tseng, C. S., Dai, L. G., & Hsu, S. H. (2016). Water-based polyurethane 3D printed scaffolds with controlled release function for customized cartilage tissue engineering. *Biomaterials*, 83, 156–168.

Ishida, Y., & Miyasaka, T. (2016). Dimensional accuracy of dental casting patterns created by 3D printers. *Dental Materials Journal*, 35(2), 250–256.

Jacobs, C. A., & Lin, A. Y. (2017). A new classification of three-dimensional printing technologies: systematic review of three-dimensional printing for patient-specific craniomaxillofacial surgery. *Plastic and Reconstructive Surgery*, 139(5), 1211–1220.

Jain, A. R., Nallaswamy, D., Ariga, P., & Ganapathy, D. M. (2018). Determination of correlation of width of Maxillary Anterior Teeth using Extraoral and Intraoral Factors in Indian Population: a systematic review. *World Journal of Dentistry*, 9(1), 68–75.

Jain, R., Supriya, B. S., & Gupta, K. (2016). Recent trends of 3-D printing in dentistry-A review. *Annals of Prosthodontics and Restorative Dentistry*, 2(1), 101–104.

Kalberer, N., Mehl, A., Schimmel, M., Müller, F., & Srinivasan, M. (2019). CAD-CAM milled versus rapidly prototyped (3D-printed) complete dentures: an in vitro evaluation of trueness. *The Journal of Prosthetic Dentistry*, 121(4), 637–643.

Kalman, L. (2018). 3D printing of a novel dental implant abutment. *Journal of Dental Research, Dental Clinics, Dental Prospects*, 12.4, 299.

Kessler, A., Hickel, R., & Reymus, M. (2020a). 3D printing in dentistry—State of the art. *Operative Dentistry*, 45(1), 30–40.

Kessler, A., Hickel, R., & Reymus, M. (2020b). 3D printing in dentistry—State of the art. *Operative Dentistry*, 45(1), 30–40.

Khorsandi, D., Fahimipour, A., Abasian, P., Saber, S. S., Seyedi, M., Ghanavati, S., ... & Makvandi, P. (2021). 3D and 4D printing in dentistry and maxillofacial surgery: printing techniques, materials, and applications. *Acta Biomaterialia*, 122, 26–49.

Kumar, N., Kumar, H., & Khurmi, J. S. (2016). Experimental investigation of process parameters for rapid prototyping technique (selective laser sintering) to enhance the part quality of prototype by Taguchi method. *Procedia Technology*, 23, 352–360.

Largo, R. D., & Garvey, P. B. (2018). Updates in head and neck reconstruction. *Plastic and Reconstructive Surgery*, 141(2), 271e–285e.

Lin, A. Y., & Yarholar, L. M. (2020). Plastic surgery innovation with 3D printing for craniomaxillofacial operations. *Missouri Medicine*, 117(2), 136.

Louvrier, A., Marty, P., Barrabé, A., Euvrard, E., Chatelain, B., Weber, E., & Meyer, C. (2017). How useful is 3D printing in maxillofacial surgery?. *Journal of Stomatology, Oral and Maxillofacial Surgery*, 118(4), 206–212.

Mai, H. N., Lee, K. B., & Lee, D. H. (2017). Fit of interim crowns fabricated using photopolymer-jetting 3D printing. *The Journal of Prosthetic Dentistry*, 118(2), 208–215.

Mangano, C., Bianchi, A., Mangano, F. G., Dana, J., Colombo, M., Solop, I., & Admakin, O. (2020). Custom-made 3D printed subperiosteal titanium implants for the prosthetic restoration of the atrophic posterior mandible of elderly patients: a case series. *3D Printing in Medicine, 6*(1), 1–14.

McLaughlin, J. B., Ramos Jr, V., & Dickinson, D. P. (2019). Comparison of fit of dentures fabricated by traditional techniques versus CAD/CAM technology. *Journal of Prosthodontics, 28*(4), 428–435.

Negm, E. E., Aboutaleb, F. A., & Alam-Eldein, A. M. (2019). Virtual evaluation of the accuracy of fit and trueness in maxillary poly (etheretherketone) removable partial denture frameworks fabricated by direct and indirect CAD/CAM techniques. *Journal of Prosthodontics, 28*(7), 804–810.

Negruțiu, C. S. (2017). Digital dentistry—digital impression and CAD/CAM system applications. *Journal of Interdisciplinary Medicine, 2*(1), 54–57.

Nesic, D., Schaefer, B. M., Sun, Y., Saulacic, N., & Sailer, I. (2020). 3D printing approach in dentistry: the future for personalized oral soft tissue regeneration. *Journal of Clinical Medicine, 9*(7), 2238.

Osman, R. B., van der Veen, A. J., Huiberts, D., Wismeijer, D., & Alharbi, N. (2017). 3D-printing zirconia implants; a dream or a reality? An in-vitro study evaluating the dimensional accuracy, surface topography and mechanical properties of printed zirconia implant and discs. *Journal of the Mechanical Behavior of Biomedical Materials, 75*, 521–528.

Papadiochou, S., & Pissiotis, A. L. (2018). Marginal adaptation and CAD-CAM technology: A systematic review of restorative material and fabrication techniques. *The Journal of Prosthetic Dentistry, 119*(4), 545–551.

Papaspyridakos, P., Gallucci, G. O., Chen, C. J., Hanssen, S., Naert, I., & Vandenberghe, B. (2016). Digital versus conventional implant impressions for edentulous patients: accuracy outcomes. *Clinical Oral Implants Research, 27*(4), 465–472.

Patzelt, S. B., Vonau, S., Stampf, S., & Att, W. (2013). Assessing the feasibility and accuracy of digitizing edentulous jaws. *The Journal of the American Dental Association, 144*(8), 914–920.

Paudyal, M. (2015). A Brief Study on Three-Dimensional Printing Focusing on the Process of Fused Deposition Modeling.

Prechtel, A., Stawarczyk, B., Hickel, R., Edelhoff, D., & Reymus, M. (2020). Fracture load of 3D printed PEEK inlays compared with milled ones, direct resin composite fillings, and sound teeth. *Clinical Oral Investigations, 24*(10), 3457–3466.

Presotto, A. G. C., Barão, V. A. R., Bhering, C. L. B., & Mesquita, M. F. (2019). Dimensional precision of implant-supported frameworks fabricated by 3D printing. *The Journal of Prosthetic Dentistry, 122*(1), 38–45.

Punia, U., Kaushik, A., Garg, R. K., Chhabra, D., & Sharma, A. (2022). 3D printable biomaterials for dental restoration: A systematic review. *Materials Today: Proceedings*.

Raza, S. M., & Singh, D. (2020). Experimental Investigation on Filament Extrusion using recycled materials.

Sandeep, D. C., & Chhabra, D. (2017). Comparison and analysis of different 3d printing techniques. *International Journal of Latest Trends in Engineering and Technology, 8*, 264–272.

Saponaro, P. C., Yilmaz, B., Heshmati, R. H., & McGlumphy, E. A. (2016). Clinical performance of CAD-CAM-fabricated complete dentures: a cross-sectional study. *The Journal of Prosthetic Dentistry*, 116(3), 431–435.

Sawhney, H., & Jose, A. A. (2018). 3D Printing in Dentistry—Sculpting the Way It Is. *Journal of Scientific and Technical Research*, 8, 1–4.

Schweiger, J., Edelhoff, D., & Güth, J. F. (2021). 3D printing in digital prosthetic dentistry: an overview of recent developments in additive manufacturing. *Journal of Clinical Medicine*, 10(9), 2010.

Sharma, A., Chhabra, D., Sahdev, R., Kaushik, A., & Punia, U. (2022). Investigation of wear rate of FDM printed TPU, ASA and multi-material parts using heuristic GANN tool. *Materials Today: Proceedings*.

Sheela, U. B., Usha, P. G., Joseph, M. M., Melo, J. S., Nair, S. T. T., & Tripathi, A. (2021). 3D printing in dental implants. In *3D Printing in Medicine and Surgery* (pp. 83–104). Woodhead Publishing.

Sikder, P., Ferreira, J. A., Fakhrabadi, E. A., Kantorski, K. Z., Liberatore, M. W., Bottino, M. C., & Bhaduri, S. B. (2020). Bioactive amorphous magnesium phosphate-polyetheretherketone composite filaments for 3D printing. *Dental Materials*, 36(7), 865–883.

Sood, A. K. (2011). *Study on Parametric Optimization of Fused Deposition Modelling (FDM) Process* (Doctoral dissertation).

Tedesco, J., Lee, B. E., Lin, A. Y., Binkley, D. M., Delaney, K. H., Kwiecien, J. M., & Grandfield, K. (2017). Osseointegration of a 3D printed stemmed titanium dental implant: A pilot study. *International Journal of Dentistry*, 2017, 1–11.

Telfer, S., & Woodburn, J. (2010). The use of 3D surface scanning for the measurement and assessment of the human foot. *Journal of Foot and Ankle Research*, 3(1), 1–9.

Van Noort, R. (2012). The future of dental devices is digital. *Dental Materials*, 28(1), 3–12.

Vashistha, R., Kumar, P., Dangi, A. K., Sharma, N., Chhabra, D., & Shukla, P. (2019). Quest for cardiovascular interventions: precise modeling and 3D printing of heart valves. *Journal of Biological Engineering*, 13(1), 1–12.

Wang, W., Yu, H., Liu, Y., Jiang, X., & Gao, B. (2019). Trueness analysis of zirconia crowns fabricated with 3-dimensional printing. *The Journal of Prosthetic Dentistry*, 121(2), 285–291.

Wang, X., Jiang, M., Zhou, Z., Gou, J., & Hui, D. (2017). 3D printing of polymer matrix composites: A review and prospective. *Composites Part B: Engineering*, 110, 442–458.

Wilk, B. L. (2015). Intraoral digital impressioning for dental implant restorations versus traditional implant impression techniques. *Compendium of Continuing Education in Dentistry*, 36, 529–533.

Wu, J., Cheng, Y., Gao, B., & Yu, H. (2020). A novel digital altered cast impression technique for fabricating a removable partial denture with a distal extension. *The Journal of the American Dental Association*, 151(4), 297–302.

Yildirim, B. (2020). Effect of porcelain firing and cementation on the marginal fit of implant-supported metal-ceramic restorations fabricated by additive or subtractive manufacturing methods. *The Journal of Prosthetic Dentistry*, 124(4), 476-e1.

Zaharia, C., Gabor, A. G., Gavrilovici, A., Stan, A. T., Idorasi, L., Sinescu, C., & Negruţiu, M. L. (2017). Digital dentistry-3D printing applications. *Journal of Interdisciplinary Medicine*, 2(1), 50–53.

Chapter 10

Dental Brace Development Using Digital Light Processing 3D Printing

W.F. Tang, C.C. Lee, S.L. Mak, C.H. Li, and Eddy Chan

CONTENTS

10.1 INTRODUCTION

The popularity of three-dimensional (3D) printing technology, also known as additive manufacturing technology, is accelerating. Not only is rapid prototyping a significant component of the mainstream adoption of additive manufacturing technology, but also is the rising emphasis on improving

DOI: 10.1201/9781003306238-10

printing quality and generating higher-resolution products with complicated geometry. Fused deposition modeling (FDM), one of the 3D printing processes, was primarily utilized mostly for rapid tooling and prototyping, allowing engineers to enhance the design of basic tools and prototypes. Additive manufacturing technology is comprehensive enough to allow professionals in a variety of industries to build customized products in complicated geometries. Since the production of dental braces requires precise and complicated structures, this trend has resulted in growing of research on additive manufacturing applications in the dental industry. Dentists may use this technology to build customized transparent plastic braces to replace conventional metal braces for orthodontic correction in a short period of time by combining digital scanning and 3D printing techniques. Aside from the issue of "visible" stainless steel in the mouth, metal braces are not removable and cause food limitations, raising questions about their use. As a result, there is a need to develop a comprehensive solution for replacing conventional metal braces with translucent plastic braces for orthodontic correction. Because FDM is no longer suitable for building complicated plastic objects, a practical 3D printing approach for building high-precision products is required. Digital light processing (DLP) is a new additive manufacturing technology that is being utilized to create smaller, more detailed items. The study's objectives are to (1) examine 3D scanning technologies; (2) identify the critical process parameters for fabricating transparent plastic braces utilizing Digital light processing (DLP) technology; and (3) identify plastic brace materials with sufficient resilience for orthodontic correction.

With the aims of enhancing dental health, a French dentist Pierre Fauchard established the first traditional metal brace in 1728. Metal brace wires do have quite a lot of restrictions, though. The mouth may become pole and irritated. Additionally, it is necessary to have brushing and flossing. As such, patients with metal braces are subject to numerous food restrictions. Regarding the time of treatment. treatment with a metal brace takes between two and five years. As a result, the invention of invisible dental braces were established.

The invisible dental brace is plastic and translucent. In comparison to a conventional metal brace, it is softer and more pleasant. Patients may eat whatever they want while using invisible braces since they can be removed for simple cleaning. An invisible brace, however, takes four to five weeks to process. The patient must visit dentist to develop a treatment plan before starting to scan their teeth and customize the mold. The scanned file is submitted to the manufacturers to be developed into a set of dental braces when the dentist accepts the treatment plan. The finalized brace is returned to the dentist. The use of one-stop production can enhance the manufacturing of invisible braces. The development of the braces within the dentist office allows for one-stop manufacture, which saves both time and money. Eliminating the need for manufacturers reduces freight and labor costs. The processing period can also be shortened to two days. Using digital light processing (DLP), it is possible to print intricate 3-D forms. It is a reasonably inexpensive method that dentist might use to produce the invisible braces for their patients.

10.2 LITERATURE REVIEW

10.2.1 The history of orthodontic aligner

The first tooth aligner was invented by an American orthodontist Harold Kesling in 1946; it was used for orthodontic treatment and was constructed by polymer. By constructing the cast model, he discovered that sizable motions could be accomplished. The tooth aligner was then constructed based on the position of teeth. Kesling stated that although the tooth aligner may provide alignment, the current treatment does not appear to be feasible in daily life. As a result, since 1946, the clear aligner has continued to be improved. In 1964, Dr. Herny Nahoum then began to extend Kesling's concept and developed the vacuum-based forming the dental aligner by utilizing the thermoplastic sheet.

An Essix aligner system was constructed by American orthodontist John J. Sheridan in 1993 and gained popularity in North America. Then he creates a number of thermoplastic overlays for use in orthodontic devices. Afterward, two MBA students established a business proposal for what would eventually become Invisalign in 1997. Clinical validation was developed by Sheridan and the aligner technology was developed and popular. Dental aligners have been utilized by more than two million patients globally. With the advancements in software and materials, there are several sophisticated manufacturing processes now available.

10.2.2 Additive manufacturing technologies

The technology of DLP projects the illuminations in form of a single image in each layer on the printing platform. Each layer is made up of tiny rectangular squares known as pixels since the projected picture is a digital screen and each layer's image is composed of square pixels (Lombardo et al., 2017). The 3D model file is imported to the 3D printer and the printing process is launched.

Reddy claims that given safelight circumstances, a DLP projector transfers the image of a 3D model onto a vat of resin (Van Noort, 2012). The construction stage moves down as more resin is exposed to the light and hardens owing to the light sensitivity of the exposed resin. Layer by layer, the procedure is continued until the 3D model is finished. Reddy claims that the model solidifies after the resin vat is drained (Van Noort, 2012). High precision is used in the printing process.

10.2.3 Additive manufacturing process

In 2015, the *British Dental Journal* reported that the 3D printing process had five phases (Dawood et al., 2015). The first step is to develop a digital or physical 3D dental mold of the patient, which the oral scanner can perform. The 3D model's development and preparation of its essential support structure using CAD software comes next. The model is then printed using a 3D printer. Post-processing, which includes removing the supporting materials, is the final phase. The dental mold is printed out utilizing additive manufacturing technology before the fabrication by vacuum thermoforming (Kruth et al., 2005).

10.2.4 Thermoforming process

Thermoforming process is used in the final process of dental brace production. A vacuum thermoformer equipment will be used for invisible brace formation which will heat up the plastic sheet and fabricate the dental brace. Below is the procedure for vacuum fabrication process.

  a. Heat up the thermoformer under vacuum.
  b. Insert the plastic sheet into the thermoformer's sheet retaining device.
  c. Set the polymer mold on the apparatus's base plate.
  d. To heat the plastic sheet, move the heating element over it.
  e. Remove the heating element from the plastic sheet once it has warmed up and sagged.
  f. Switch on the thermoformer's vacuum system, then forcefully lower the holding fixture for the sheet over the model.
  g. To ensure a good vacuum, make sure the holding fixture is completely in touch with the model.
  h. Take the thermoformed model out of the machinery and finish trimming and polishing.

10.2.5 Pros and cons of digital light processing

The exceptional laying precision of the DLP process is one of its benefits. A pair of plastic braces may be placed into the patient's teeth since the mold created by the DLP approach will be accurate to the condition of a human's oral cavity. Additionally, the mold with a smooth printing surface produced using the DLP technology may assist us in creating a plastic brace with a high level of finish, which will affect the brace's functionality (Wu et al., 2017). The production cost will also be influenced and affected by the low cost of DLP printers. However, the printing substance is a unique resin that is light-sensitive. The high material cost may contribute in a rise in treatment costs.

10.2.6 Finishing of plastic aligner

Prior to application, the brace's edge and surface should be smoothed to lower the chance of a sharp cut and improve comfort. The dental brace trimming and finishing processes and methods were offered by Zendura Dental (2019). The method it suggests is listed below.

a. Using cutters or trimmers to remove extra brace material from the dental mold.
b. Take the brace out of the mold. Insert a little straight screwdriver where indicated and flip the upside-down brace that was created using the dental mold. To get the brace out of the mold, gently twist the screwdriver. More deeply insert the tip of the screwdriver into the gap between the produced sheet and the model. The brace is taken apart.
c. Once more, trim the brace's edge-to-edge extra material. Cut between 1.0 and 1.2 cm beneath the gingiva line.
d. Smoothen the cut edges of the brace with polish. Establish a medium grit by combining a coarse grit with a rotary tool and polishing wheel.
e. Using water to rinse the brace.

10.3 METHODOLOGY

10.3.1 Fourier-transform infrared spectroscopy

Fourier-transform infrared spectroscopy (FTIR) test is conducted to determine the composed material of thermoplastic sheets. We will use two different thermoplastic sheets; one is a soft sheet while the other one is a splint sheet. Two samples of each thermoplastic sheets will be tested. Below shows the workflow (Figure 10.1) of FTIR test.

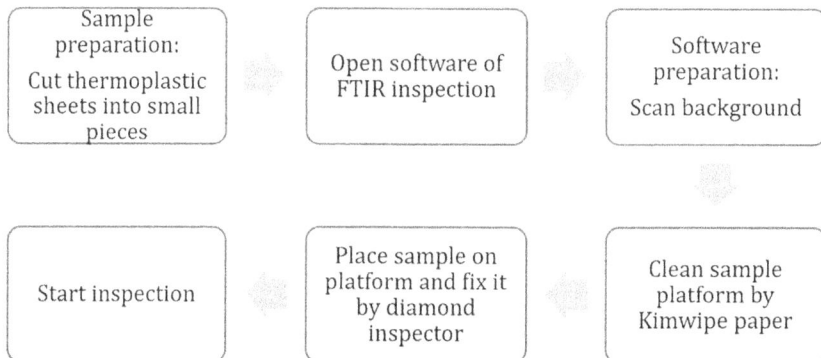

Figure 10.1 FTIR measurement workflow.

The composition of thermoplastic sheets is determined using Fourier-transform infrared spectroscopy. Two types of thermoplastic sheets were evaluated, one of which is soft and the other of which is splint. The FTIR test procedure is illustrated in Figure 10.1.

10.3.2 Hardness test

Two different types of thermoplastic sheets with a thickness of 1.0 mm were tested for hardness using the Shore D hardness scale. The test specimen must meet the requirements of ASTM D2240. Thickness requirement of test samples should be 6.0 mm. The thermoplastic sheets are divided into six pieces and placed on top of one another to achieve the desired thickness of 6.0 mm. During the hardness measurement, hardness value on a scale of 0 to 100 used to assess hardness.

10.3.3 Dimension and profile measurement

As illustrated in Figure 10.2, two horizontal distances between the second molars—tooth 17 to tooth 27 in the upper jaw and tooth 47 to tooth 37 in the lower jaw—were measured. This is to determine the accuracy of the size of actual mode in relation to the scanning profile.

Three times of printing are made throughout the printing process. Both the top jaw and lower jaw are printed individually for the first and second printings. Daylight Hard Resin is used. For the third printing, both the upper and lower jaws are produced simultaneously and Daylight Firm Resin was utilized. The printing process took around seven hours. All three prints have the same layer height of 50 microns. Table 10.1 illustrate the dimension measurement of upper jaw and lower jaw for the following three: (1) brace; (2) printed mold; and (3) scanning software.

The procedure of vacuum thermoforming is then used to develop a set of dental braces by using both soft and splint sheets. Following vacuum forming, excessive thermoplastic sheet residues was cut out and removed using a trimming procedure. The production procedure for dental braces is seen in Figure 10.3.

10.4 RESULTS

10.4.1 FTIR

The material qualities of the sheets utilized for the soft and splint samples are verified using FTIR. Ethylene terephthalate makes up the majority of the sample sheets' components (Figures 10.4–10.7).

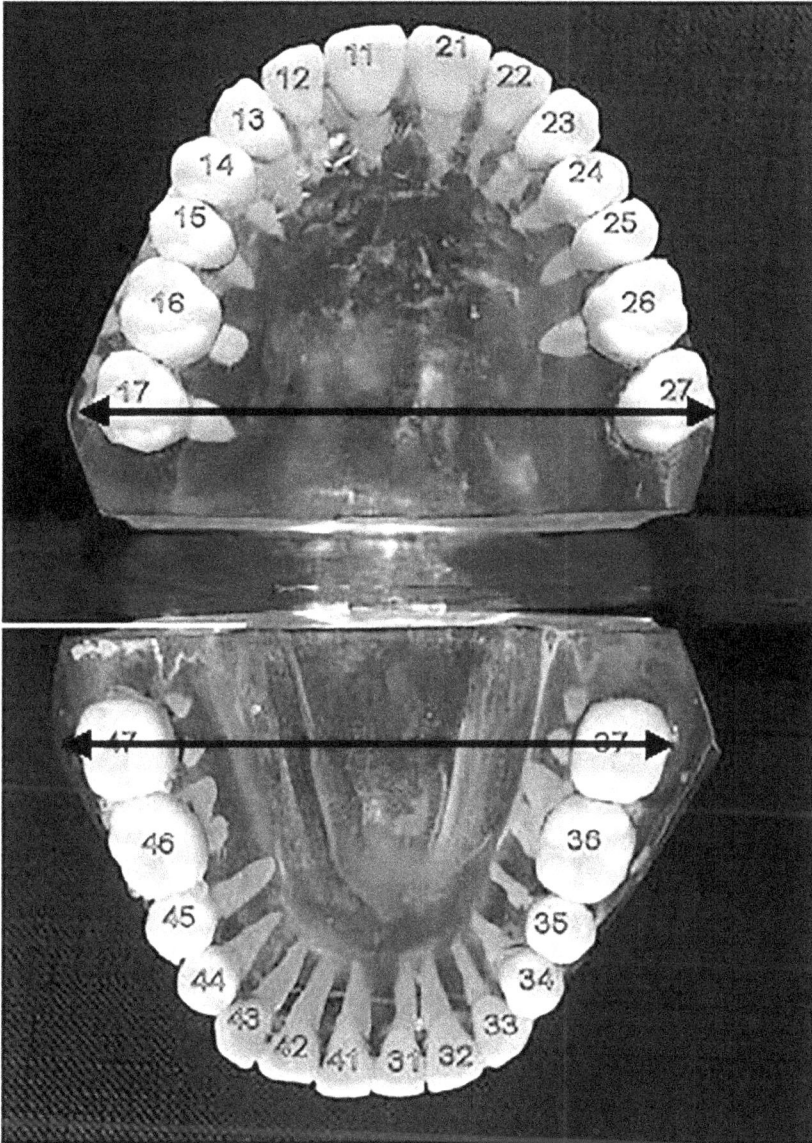

Figure 10.2 Dimension measurement on dental brace.

10.4.2 Thermoforming

It is noted that the performance of thermoplastic sheets used to make dental braces varies. Since it adhered to the dental mold, a pair of dental braces made from soft sheet is difficult to remove. On the other hand, a different set of dental braces made from splint sheet does not adhere to

Table 10.1 Dimension measurement of upper and lower jaw

	Distance (mm)	Difference|(mm)
Upper Jaw		
Brace (splint)	62.14	0.872 (with printed mold)
Printed mold	61.268	0.016 (with software)
Scanning software	61.252	N/A
Lower Jaw		
Brace (splint)	57.1	0.164 (with printed mold)
Printed mold	56.936	0.04 (with software)
Scanning software	56.976	N/A

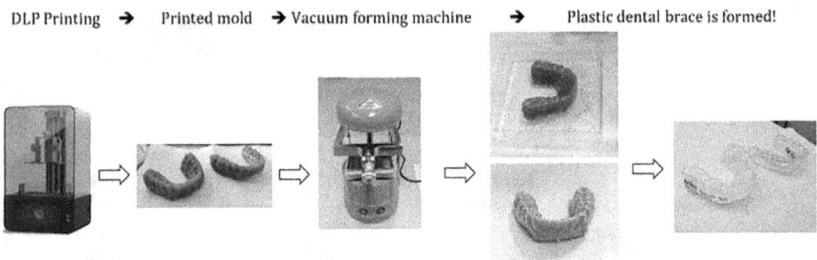

DLP Printing → Printed mold → Vacuum forming machine → Plastic dental brace is formed!

Figure 10.3 Manufacturing process of the invisible dental brace.

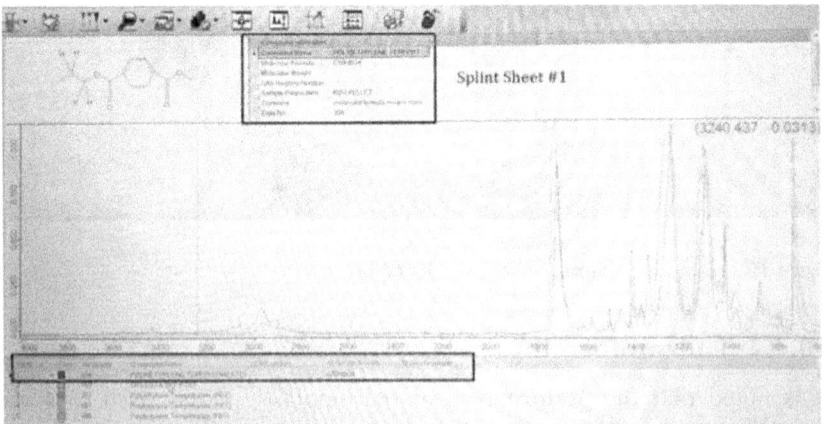

Splint Sheet #1

Figure 10.4 FTIR results for soft and splint sheets.

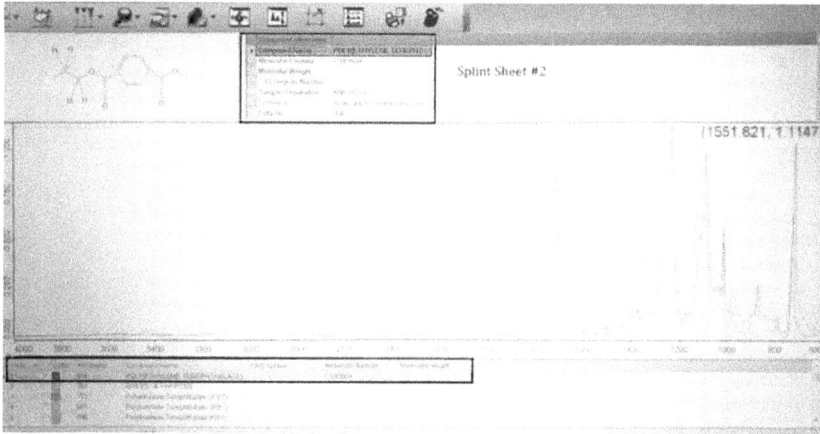

Figure 10.5 FTIR results for soft and splint sheets.

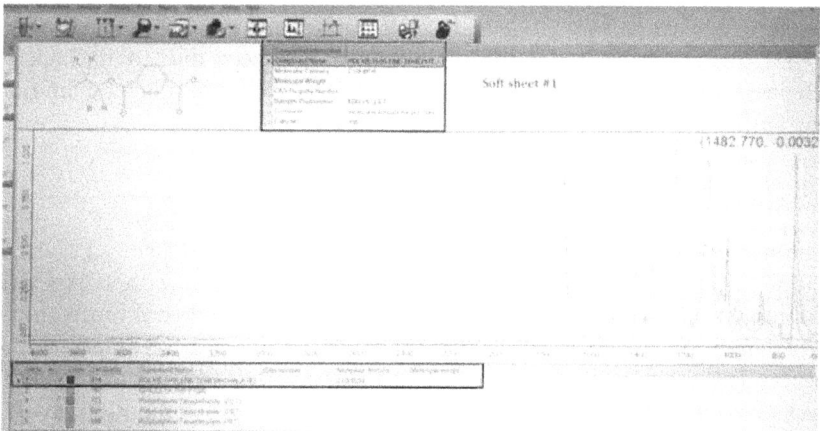

Figure 10.6 FTIR results for soft and splint sheets.

the dental mold but requires a longer heating period to soften the thermoplastic sheet.

10.4.3 Hardness measurement

Al Noor& Al-Joubori (2018a,b) stated that the hardness of the aligners has a significant impact on the elasticity. As such, the hardness of the dental brace sheet materials was measured. The average hardness of the splint sheet is 72 HRD, which is greater than the average hardness of the soft sheet, which is 26 HRD, after the hardness of the two materials was measured using a Shore D Hardness Tester Dental braces with low hardness

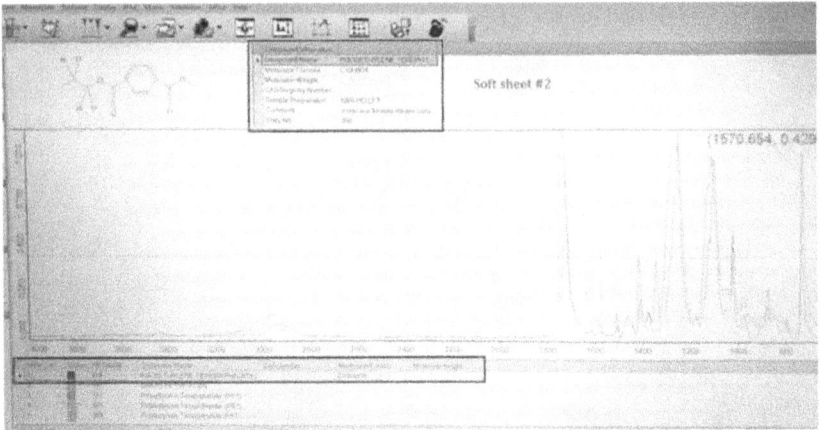

Figure 10.7 FTIR results for soft and splint sheets.

with soft sheet cannot give enough force for the teeth since they must deliver a consistent force to limit teeth from moving. As such, the soft sheet is not a suitable material for making dental braces due to its lack of hardness in comparison to the hardness of orthodontic aligner materials currently on the market.

10.4.4 Dimensional analysis

The upper jaw and lower jaw distances were measured using a Vernier caliper. According to the aforementioned finding, there is a 0.164 mm discrepancy between the lower jaw mold and the lower dental brace, and a 0.872 mm difference between the upper jaw mold and the upper dental brace (Table 10.1). The top jaw's discrepancy is 0.016 mm and the lower jaw's is 0.04 mm when comparing the printed mold's dimensions to the scanned profile. It is noted that there is less variation between the prescribed and actual dimensions. Thus, dental mold is proved to be feasible to be printed by DLP in good quality, with small dimension discrepancy between the printed mold and the scanned profile.

10.5 DISCUSSION

10.5.1 Fabrication process

The fabrication of dental braces and the material properties employed are the two key components. In terms of the production method, DLP is used to construct the upper and lower jaw resin teeth molds, and PET sheet (0.1 mm soft sheet and 0.1 mm splint sheet) is utilized to construct the dental braces. The composition of dental braces is realized using FTIR, and the hardness of PET sheet is studied using a hardness tester. The dimension measurement is then analyzed to verify if there is a substantial variation after different steps. The results indicated that PET sheet in 1.0 mm thick is acceptable for making dental braces. To deliver high accuracy, the following variables need to be accurately tracked.

10.5.1.1 The scanning device's precision and accuracy

The method that is followed will be directly impacted by the scanning device. The mold of the dental brace and the finished product will both be influenced by incorrect data input into the CAD software if the data from tooth alignmentsis incorrect (Venkatesh & Nandini, 2013).

10.5.1.2 DLP printer precision and accuracy

The resolution should be sufficient and accurate enough for usage because the quality of tooth aligner is highly affected by the DLP printer resolution (Miyazaki & Hotta, 2011).

10.5.1.3 Resin

Dental mold printing resin should be heat-resistant and should not distort at high temperatures in order to ensure accuracy and precision of the finished tooth aligners mold must withstand high temperature throughout the vacuum manufacturing process.

The materials used for dental braces' should not be easily affected by heat, therefore they should be heat resistant during the vacuum fabrication process.

10.5.2 Evaluation of 3D scanners

There are two critical methods for digitalizing dental casts or impressions: direct digitalization using an intraoral scanner, and indirect digitalization using a conventional type of desktop scanner. An appropriate scanner is recommended by evaluating their processes, typical scanning times, and accuracy. Traditional scanners would need more stages in the clinical flow method than the intraoral scanner. For the traditional scanner, the dentist will first take a patient's dental imprint before sending it to the lab to construct a plaster mold using the patient's impression. Finally, the plaster mold is scanned using the desktop scanner. Procedures for an intraoral scanner are easier and more direct than those for a conventional scanner. The oral cavity of patient's teeth can be immediately scanned. Table 10.2 illustrate the comparison on both conventional and intraoral scanning methods.

Second, a conventional scanner takes more time than an intraoral scanner to complete a scan on average of two hours, whereas an intraoral

Table 10.2 Comparison on both conventional and intraoral scanning methods

Type of 3D scanner	Procedure (clinical flow)
Conventional scanner (Desktop/ Lab type)	1. Take dental impression from the patient 2. Make a plaster cast from the impression 3. Scan the plaster cast by desktop scanner
Intraoral scanner	Scan the full-arch teeth directly from the patient's mouth

scanner only takes 20 minutes. The intraoral scanner saves additional time.

Finally, desktop scanners are more accurate than intraoral scanners in terms of precision. Wesemann, et al. (2016) come to the conclusion that intraoral scanners are very suited for manufacturing the orthodontic appliance in a fully digital process, but desktop scanners still deliver a comparable outcome. On the other hand, the aforementioned assertion is also supported by the product specifications of 3Shape, an international company that sells dental scanners.

10.5.3 Printing resin comparison

Two distinct printing resins were employed for DLP 3D printing in the research study—Daylight firm resin and Hard daylight resin. In the research study, anoptimal type of resin would be determined between these two types of resin by producing a set of dental molds. By comparing the two types of resin. Daylight firm resin is suitable to make hard molds due to itsless compressive ability under high force and minimum shrinking ability. The hard mold using the Daylight firm resin will gently flex and stretch. Because of its low elongation and high tensile strength, the hard mold can only have slightlybending and give less deformation. The curing condition is 30 minutes in UV light or 60 minutes under a typical 60W table lamp. The hard mold should be maintained in dry conditions away from UV light; it will be remained as tough, sturdy, and long lasting.

Hard daylight polymers, which have little compressive ability under strong force, give modest shrinking, and are suitable for printing hard mold. Under pressure, a printed item will not flex or deflect. Additionally, they will exhibit essentially minimal elongation and strong tensile shear characteristics. Fast exposure durations and a broad exposure latitude may be accomplished, much like with daylight firm resin. The item can stay strong, resilient, and long-lasting if it can be stored in dry settings away from UV rays.

As such, a resin with a high tensile modulus and low elongation will often be used because the dental mold cannot bend or stretch under load. Daylight Hard resin should be the best option for printing dental molds.

10.6 CONCLUSION

A good dental braces should be simple to take off and put on, fit your size, be extremely rigid, and be pleasant to wear. Reliability testing and field tests are essential for assessing the braces' performance. It is necessary to undertake further research on the mechanical and physical characteristics of materials. The materials used to make the orthodontic aligners should be examined for stiffness of thermoplastic sheet. Some researchers showed that in order to achieve tooth movement, thermoplastic materials have to be able to generate force and hold it via material deflection. The stiffness of the material determines how much deflection occurs, therefore the elastic modulus may be a good predictor of stiffness. The stiffness of the material increases with increasing elastic modulus, producing sufficient force for more accurate tooth movement. As a result, thermoplastic sheet should undergo tensile testing in accordance with ASTM D638, and elastic modulus may be determined using the stress and strain data. Additionally, it is important to study the thermoplastic material's retention force throughout a range of thicknesses. Despite the fact that there are aligner materials with varied thicknesses on the market, our project exclusively focuses on 1.0 mm thick aligner material. Studying retention force will help you choose a material thickness that is appropriate. Retention and achieving desirable tooth movement are connected. The measurement by tensile tests is to provide vertical forces that pulled the aligner off the braces. To ensure that the aligner material complies with ISO 20795-2, which outlines the specifications for orthodontic base polymers used in active or passive orthodontic equipment, a three-point bend test should be conducted. Ultimate flexural strength and flexural modulus typically need to be at least 50 MPa and 1500 MPa, respectively, according to the clauses 5.2.6 and 5.2.7 of ISO 20795-2. Field testing is advised in the future study. To simulatethe actual movement and usage of tooth aligner, the field study should be considered to verify the mouth movements such as smiling, eating, and chatting. The wearing experience will serve as the primary data point for evaluating how well the dental braces work. The chemical and antibacterial qualities of the material used in orthodontic alignment should also be the subject of future research.

REFERENCES

Al Noor, H. & Al-Joubori S. (2018a). Comparison of the hardness and elastic modulus of different orthodontic aligners' materials. *International Journal of Medical Research and Pharmaceutical Sciences*, 5(9), 19–25. doi: 10.5281/zenodo.1443358

Al Noor, H. & Al-Joubori S. (2018b). Retention of Different Orthodontic Aligners According to their Thickness and the Presence of Attachments. *International*

Journal of Medical Research & Health Sciences, 2018, 7(11), 115–121. Retrieved from https://www.ijmrhs.com/medical-research/retention-of-different-orthodontic-aligners-according-to-their-thickness-and-the-presence-of-attachments.pdf

Dawood, A., Marti B., Sauret-Jackson V. & Darwood A. (2015, December). 3D Printing in Dentistry. *British Dental Journal*, 219(11), 521–529. doi:10.1038/sj.bdj.2015.914.

Kruth, J.P., Vandenbroucke B., van Vaerenbergh J. & Naert I. (2005). Digital manufacturing of biocompatible metal frameworks for complex dental prostheses by means of SLS/SLM. In: *Da Silva Bartolo PJ: Virtual modeling and rapid manufacturing*. London/Taylor & Francis Group. 10.

Lombardo, L., Martines E., Mazzanti V., Arreghini A., Mollica F. & Siciliani G. (2017). Stress relaxation properties of four orthodontic aligner materials: A 24-hour in vitro study. *Angle Orthodontist*, 87(1), 11–18. 10.2319/113015-813.1

Miyazaki, T. & Hotta Y. (2011). CAD/CAM systems available for the fabrication of crown and bridge restorations. *Australian Dental Journal*, 56, 97–106. 7.

Trimming techniques for Zendura appliances - ZenduraDental. (n.d.). Retrieved September 23, 2019, from Zendura: https://www.zenduradental.com/pages/trimming

Vacuum Fabrication Instructions ForZendura Material - ZenduraDental. (n.d.). Retrieved September 24, 2019, Zendura: https://www.zenduradental.com/pages/vacuum-fabrication-instructions

Van Noort, R. (2012). The future of dental devices is digital. *Dental Materials*, 28, 3–12. 6.

Venkatesh, K.V. & Nandini V.V. (2013). Direct metal laser sintering: a digitised metal casting technology. *Journal of Indian Prosthodontic Society*, 13, 389–392. 8.

Wesemann, C., Muallah J., Mah J. & Bumann A. (2016). Accuracy and efficiency of full-arch digitalization and 3D printing: A comparison between desktop model scanners, an intraoral scanner, a CBCT model scan, and stereolithographic 3D printing. *Quintessence International*, 48(1). 41–50. doi: 10.3290/j.qi.a37130

Wu, J., Li Y. & Zhang Y. (2017). Use of intraoral scanning and 3Dimensional printing in the fabrication of a removable partial denture for a patient with limited mouth opening. *Journal of the American Dental Association*. pii:S0002-8177(17)30083-1. 5.

Xu, F., Wong Y.S., Loh T.H. (2000). Toward generic models for comparative evaluation and process selection in rapid prototyping and manufacturing. *Journal of Manufacturing Systems*, 19, 283–296

Yadroitsev, I., Thivillon L., Bertrand P. & Smurov I. (2007). Strategy of manufacturing components with designed internal structure by selective laser melting of metallic powder. *Applied Surface Science*, 254, 980–983. 9.

Zendura Dental (n.d.). Vacuum fabrication instructions for Zendura Material. Retrieved September 24, 2019, Zendura: https://www.zenduradental.com/pages/vacuum-fabrication-instructions

Chapter 11

Nanomaterials and 3D Printing

F. Sharifianjazi, L. Bazli, A. Esmaeilkhanian, S. Khaksar,
F. Sadeghi, and M. Reisi Nafchi

CONTENTS

DOI: 10.1201/9781003306238-11

11.1 INTRODUCTION

Human survival and evolution have depended on our ability to design tools for overcoming biological, environmental, and societal problems. However, commercially made devices have not yet reached the level of functional integration and intricacy that is frequently found in nature, like biological structures. Naturally evolved constructs often possess higher geometrical complexity, temporally and spatially varying properties, and integrated multifunctionality (Cerkvenik et al., 2017). The combination of nanomaterials and additive manufacturing, more commonly referred to as "three-dimensional printing" (3D printing), enables the fabrication of complex, functional constructs. The 3D printing of nanomaterials provides an opportunity for the production of multiscale architectures integrated with functional nanoparticles. The patterning and guidance of nanomaterial assembly during the 3D printing process can accomplish this goal (Hales et al., 2020, Reiser et al., 2019).

This level of integration with the use of conventional production techniques is difficult to be achieved without this multiscale 3D printing approach. First, current production procedures often limit the geometric complexity of structures. Photolithography, for example, is still unable to produce completely freeform structures with defined voids (Madou, 2018). Second, unlike the hierarchical assembly of a biological system, conventional manufacturing techniques frequently require a post-fabrication assembly procedure to add functional elements. To maintain modularity, considerable design tradeoffs must be made to offset high tooling and production costs. There are constraints on the degree of functional integration available with traditional production methods. Third, customizing device properties quickly using a typical production process is an arduous task. Modulation necessitates considerable design, material, and fabrication process changes, increasing lead time and development costs. For example, functional gradients in biological systems need the capacity to change functional characteristics on-the-fly during construction, which is apparently difficult to be obtained by conventional manufacturing (Zhang et al., 2018).

Incorporating nanoparticles into 3D printing methods allows for fine-tuning of useful material qualities. Nanomaterials have dimensions in the range of 1–100 nm. Because nanomaterials' functional qualities rely on their size, they can be tailored to specific applications without changing the basic materials (Zhang et al., 2018). When the particle size is less than the Bohr exciton radius, the size-dependent energy band gap is observed in semiconductive nanomaterials (like core–shell semiconductor quantum dots) (Goesmann and Feldmann, 2010). Moreover, the surface-to-volume ratio increases when the dimensions of materials are in the nanoscale range. As a result of this geometrical effect, some parameters including melting point and surface reactivity can be modulated. Recent 3D printing breakthroughs have shown multiscale integration of nanomaterials to improve electrical, mechanical, optical, biological, thermal, and actuation properties (Gissibl et al., 2016). Controlling the deposition of nanomaterials (through evaporation and electrical forces) can also contribute anisotropy and heterogeneity to 3D printed constructions' functional characteristics. It is possible to freeform fabricate highly complex, heterogeneous, functional designs using 3D printing and nanomaterials (Al-Milaji et al., 2018).

11.2 NANOFIBERS

Fiber-reinforced 3D printing that is employed to create high-performance composites has recently been a hot study topic due to its numerous potential uses in the aerospace, automotive, building, and naval industries (Kabir et al., 2020b). Material extrusion-based printing, notably fused deposition modeling (FDM), is the highly preferred technology owing to its low cost besides ease of incorporating fiber into composite structures (Chakraborty and Biswas, 2020). Continuous fiber reinforcement necessitated two different supply systems for fiber and polymer; however, through one extrusion process (coaxial extrusion), or a separate supply of polymer and polymer pre-impregnated fiber employing dual nozzle systems (Kabir et al., 2020a). The latter is more favorable due to achieving enhanced fiber–polymer interfacial adhesion and desired fiber placing facility. On the other hand, high fiber content is obtained by the coaxial extrusion configuration; however, interfacial bonding between polymer and fiber is relatively poor, resulting in poor mechanical properties and thereby premature composite failure. It is well known that the fiber content influences the mechanical properties of composites. In order to modify mechanical characteristics, increasing fiber content in the test direction improves tensile strength (Hetrick et al., 2021). For example, the incorporation of bacterial cellulose nanofibers (BCNFs) into silk fibroin/gelatin hydrogels was carried out via a plotting-based 3D-printing platform to improve physical behavior and resolution. The BCNFs had no effect on the printability of the composite ink. The tensile strength of the 3D-printed scaffolds was increased from 100 to 800 kPa by the addition of the BCNFs. Adding BCNFs to ink improves shape fidelity and prevents collapse. Raising the BCNF concentration to 0.70 wt% led to an improvement in the quality of 3D-printed grids (Huang et al., 2019).

To achieve strong adhesion between beads and neighboring print layers for increased structural integrity, 3D printing involves more matrix material compared to traditional composite production technologies, including injection molding, compression molding, as well as resin transfer molding. Furthermore, while traditional composites can often reach 60% fiber volume fraction, 3D-printed composites with 40% of fiber volume fraction are rare (Chabaud et al., 2019). While several investigations reported 3D printed composites containing more than 40% fiber volume fraction, they failed to account for the amount of matrix in the fiber filament (also known as composite filament), resulting in an overestimation of fiber content. Pre-impregnated fiber filament is employed in dual nozzle printers. A thermogravimetric analysis (TGA) of the most widely used commercial fiber filaments showed that they contain 44% or less fiber (high strength and high-temperature fiberglass 36%, carbon 44%, Kevlar 37%, and fiberglass 38%), implying that composites with 44% fiber cannot be fabricated even with the maximum number of layers. In fact, there are certain inescapable thermoplastic/polymeric layers, including a wall layer around the composite structure, a roof layer, along with a floor layer (Al Abadi et al., 2018, Araya-Calvo et al., 2018).

11.3 NANOPARTICLES

Metal nanoparticles are studied for their optical, electrochemical, and thermal characteristics. Depending on the method, metal nanoparticles can be shaped as spheres, cubes, or rods. Micro-electronics and drug delivery are among the uses of metal nanoparticles (Azadani et al., 2021, Niazvand et al., 2020). Metal nanoparticles have recently been used to improve sintering and create new 3D printing materials. A decrement of about 900 °C can occur in the metal nanoparticles' melting temperature because of the thermodynamic size effect. As an example, the melting point of bulk gold is 1063 °C can be reduced to 130 °C for a gold cluster that has a mean diameter of 2 nm (Buffat and Borel, 1976). This allows for lower sintering temperatures and better ultimate product quality. Crane et al. reported on the use of metal nanoparticles for additive manufacturing (AM) in 2006. Crane employed iron NPs in the range of 7 to 10 nm to increase the sintering quality of 3D-printed steel products (3DP). The test specimens were printed using a steel powder ranging in diameter from 63 to 90 μm and a traditional binder. Therefore, Metal NPs can improve the sintering properties of final products by reducing distortion and shrinkage. In addition, highly concentrated metal NP inks could be used to fabricate microconnectors and microelectrodes for solar cells, batteries, and microelectronics (Crane et al., 2006).

The electrical, mechanical, and thermal properties of polymers have been improved by adding carbon nanotubes (CNTs). They are used in field emission devices, hydrogen storage, nerve regeneration scaffolds, and sensors (Naseer et al., 2018). Carbon nanoparticles added to SL resins and laser sintering (LS) materials can increase mechanical characteristics. Carbon nanoparticles improve the final parts' electrical conductivity. Finally, carbon nanotube–containing tissue scaffolds improved cell proliferation rates in culture tests (Goodridge et al., 2011).

Ceramics like SiO_2 and TiO_2 with semiconducting characteristics have been applied in optoelectronic devices, bio-imaging, labeling, and solar cells. To improve the mechanical properties of printed items, ceramic and semiconductor nanoparticles have been added to SL, LS, as well as direct writing materials. Tissue scaffolds and biodegradable ceramic nanostructures are also possible to be made for bone tissue engineering (Duan et al., 2010, Yugang et al., 2011).

11.4 NANOPLATELETS/NANOSHEETS

Graphene-reinforced nanocomposites were also produced by 3D printing techniques for different applications. The viscosity of the liquid resin was increased by 14% and the fracture toughness by 28% by adding 0.5 wt% of graphene nanoplatelets to a phenylbis (2,4,6-trimethylbenzoyl)–phosphine oxide (Irgacure 819), PLA–polyurethane oligomer (PLA–PUA), and also triethylene glycol dimethacrylate ink. Graphene oxide (GO) composite was also produced by the SL technique with improved strength and ductility to develop load-bearing scaffolds for bone regeneration (Lin et al., 2015). Figure 11.1 shows the electrical and mechanical improvement of polyurethane (PU) by the addition of graphene sheets (GS) fabricated via digital light processing (DLP) (Joo and Cho, 2020).

Another study used high-strength hydrogel scaffolds that were made of poly(N-acryloyl glycinamide) and nanoclay. The scaffolds were 3D printed by an extrusion-based system followed by polymerization using UV light irradiation with the purpose of repairing tibia bone defects (Figure 11.2). It

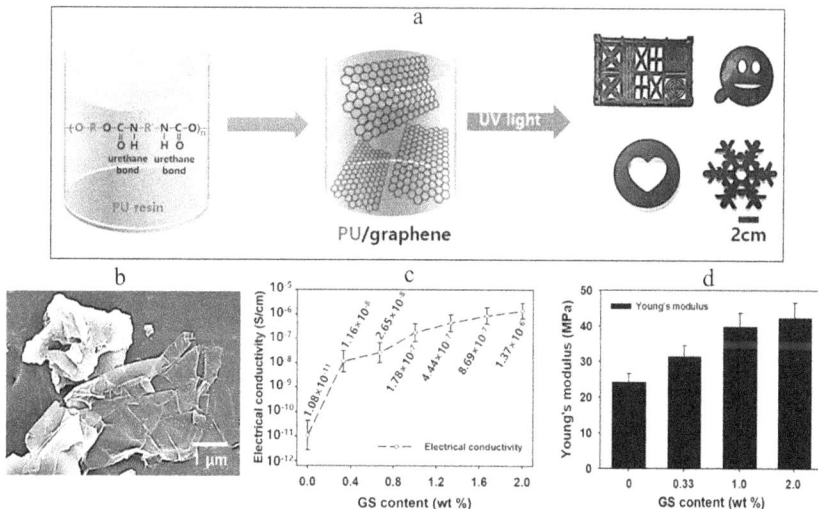

Figure 11.1 (a) Preparation, (b) SEM analysis, (c) electrical conductivity, and (d) mechanical property of PU/GS composite.

(Source: Joo and Cho 2020).

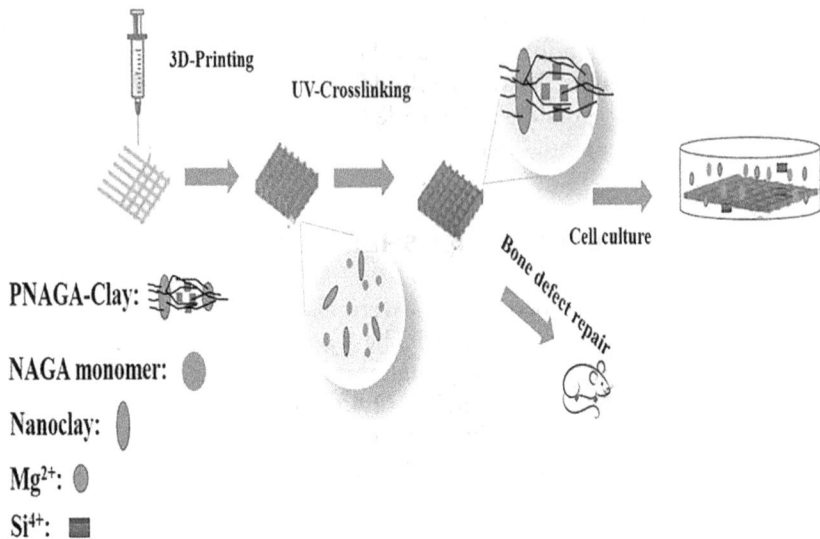

Figure 11.2 3D printed scaffolds using UV light irradiation.

was shown that the printed scaffolds' improved mechanical characteristics and magnesium (Mg^{2+}) and silicon (Si^{4+}) release aided osteogenic differentiation and thereby bone formation. It showed that nanotechnology is able to control the cellular microenvironment and microstructure of 3D-printed scaffolds, improve cell behavior, and consequently stimulate bone formation (Zhai et al., 2017).

In order to achieve a compromise between biocompatibility and printability, nanoclay was added to Gelatin methacryloyl (GelMA). The branched vessel and the ear were also printed using a screw-driven extrusion method with GelMA/nanoclay ink. Introducing nanoclay increased porosity and improved physical behavior while maintaining biocompatibility. This study demonstrated a method for 3D printing complicated GelMA-based structures with good shape integrity for bone regeneration (Gao et al., 2019).

11.5 NANO PATTERNING

11.5.1 Fluid shear patterning

In fluid shear patterning, shear stresses in flowing ink are used to control anisotropic nanomaterial orientation during 3D printing. Fluid shear patterning can easily be combined with extrusion-based printing approaches like direct ink writing (DIW) to add functionality. Using a unique process called linear harmonic oscillation (also known as chaotic printing), this approach may be expanded to bulk resin-based 3D printing. Fluid shear patterning is compatible with more materials than magnetic patterning, including silicone polymers, cellulose, hydrogels, gelatin, ceramic and metallic nanorods, as well as biological cells. A wide range of constructs can be programmed, including shape-changing structures with tunable electrical and optical characteristics. It is true that 3D printing with fluid shear patterning enables seamless integration of functional features (Yunus et al., 2016, Sydney Gladman et al., 2016).

11.5.2 Evaporative patterning

Evaporative patterning can create thin-film patterns, freestanding architectures, and high aspect-ratio structures. It is a convenient way for producing architectures with variable functional features while no extra transfer substrates or stamps are used. This is achievable by various control techniques such as binary solvents producing Marangoni flow, meniscus-guided printing, substrate alteration, 3D self-shaping templates, along with coprinting of wetting droplets and supporting. Direct writing of a colloidal solution allows the creation of freestanding structures by directing the growth of colloidal structures relative to the rate of evaporation. This patterning process is attractive for fabricating objects with tunable optical functionality (Tan et al., 2018). Understanding soft matter physics and multiphase dynamics, including drying of colloidal fluids and evaporation kinetics, is required. Understanding evaporation kinetics and colloidal drying can help control nanomaterial morphology and assembly, ultimately leading to useful nanodevices. Current research has produced analytical models, numerical simulations, and varied structures, but not macroscale, defect-free structures. Defects induced by phenomena like the coffee-ring effect and the nanomaterial assembly aggregation limit this. The resolution of 3D printing technology also limits the ability to build hierarchical systems found in nature (Zhang et al., 2017).

11.5.3 Acoustic patterning

Acoustic field-assisted patterning may be a viable way for assembling nanoparticles in 3D-printed structures. This method can create patterns ranging from simple unidirectional nanomaterial alignment to sophisticated heterogeneous patterns where particles are organized to produce microstructures having anisotropic functional characteristics. In 3D printing, direct substrate vibration or acoustic tweezers produce a patterned waveform. To modify particle sizes in a particular pressure range, one must first understand the complicated interactions between particles, fluid, and acoustic energy (Greenhall et al., 2016). Due to the limited attenuation of ultrasonic waves in fluids, ultrahigh field strengths are not required. In addition to electrical conductivity, acoustic field-assisted patterning has improved mechanical strength. While recent research has shown the ability to simulate nanoparticle assembly and build functioning devices, a uniform approach is still needed. This standard approach could help fabricate multifunctional devices with nanomaterial composition and variable functional features (Reyes et al., 2018).

11.5.4 Electrical patterning

In electrical patterning, electric fields are used to shape extruded ink and pattern nanoparticles. This process is utilized for producing patterned nanomaterials such as nanoscale droplets and fibers or align nanomaterial reinforcement (Shin et al., 2019). To create functional structures with controlled mechanical, electrical, and optical properties, and precise microstructures of well-aligned nanofiber grids, electrical patterning is used. Electrohydrodynamic (EHD) printing uses the charged state of the fluid jet for direct patterning of nanodroplets and nanofibers from a liquid medium, which results in high-resolution, multimaterial 3D constructions. Parallel electrode plates are also used to create horizontal electric fields across bulk printing processes for regulating nanomaterials' orientation. Electrical patterning creates objects with highly parallel nanofiber or nanotube/nanoplatelet alignment. Further reading on nanoscale construct manufacturing utilizing electrical forces is recommended (Chavez et al., 2019).

11.5.5 Magnetic patterning

With magnetic patterning, one can control the alignment and distribution of magnetic nanoparticles. In this process, anisotropic particles are aligned with a magnetic field, allowing multiscale, heterogeneous composites to be programmed. Biaxial alignment in 2D nanomaterials can be programmed using high-frequency rotating magnetic fields and pinning both principal axes in the rotation plane. Also, by making a balance between magnetic dipole attraction and steric and electrostatic repulsion, the distance between magnetic particles can be controlled. Additional magnetic nanomaterial placement can create smart structures with actuation and diverse composition. Magnetic patterning can fine-tune magnetic domain orientations, optical anisotropy, and mechanical properties, allowing for magnetically programmable structures and greater customization of 3D-printed electronics (Yim et al., 2013, Yim and Sitti, 2012).

11.5.6 Optical patterning

To create nanoscale and hierarchical structures, optical patterning uses photonic power to photopolymerize resin. Optical trapping, microstereolithography, and two-photon polymerization are used to achieve this. Using nanomaterials improves the spectrum of optical patterning, which allows light-based 3D printing of structures with micro- to nano-scale features (Li et al., 2019). Because of the precise control of photon emission, optical trapping's capacity to create nanostructures at the resolution of a

single particle, and developments in multiscale production, optical patterning is practiced as a promising nanofabrication technology. Laser-
printed micro-optics have low optical aberrations due to their high surface
quality. Printed metamaterials with 100 nm precision unit cells can have
tunable mechanical characteristics. However, larger print sizes, better
selection of materials, as well as higher throughput are required for expanding its uses (Bückmann et al., 2012).

11.5.7 Thermal patterning

Thermal patterning employs thermal energy for regulating the processing of
heat-sensitive materials like phase shifts, and pyroelectric effects, along with
particle interactions. Thermal patterning has limitations in some applications
involving temperature-sensitive specimens, for example, cell-infused tissue
regeneration scaffolds. On the other hand, thermal patterning can be used to
effectively generate microstructures and modulate nanomaterial patterns.
Additionally, it has compatibility with an extensive range of materials and is
able to easily integrate into a variety of extrusion and bulk 3D printing
techniques (Boley et al., 2017). Figure 11.3 depicts the physical phenomena
that are used for nano-patterning.

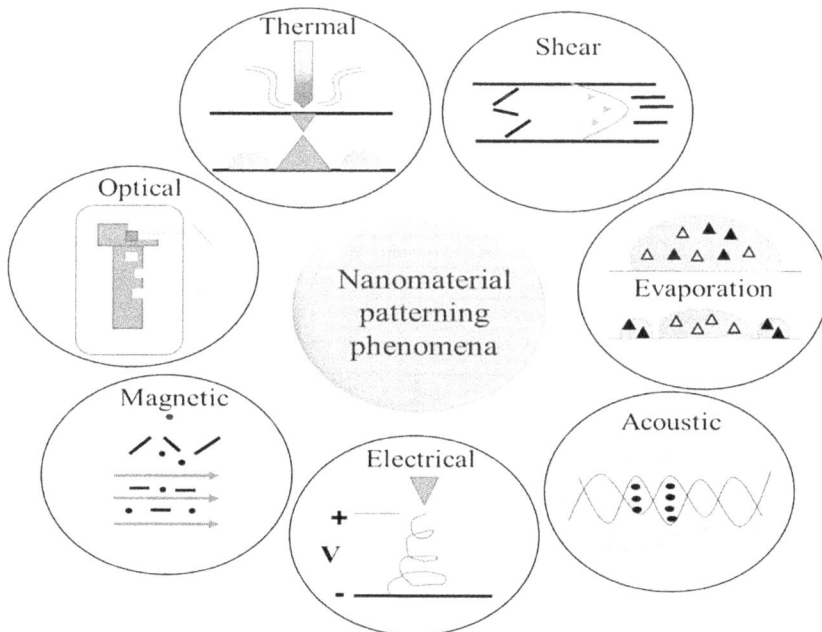

Figure 11.3 Physical phenomena used for nano-patterning.

11.6 3D-PRINTED DEVICES WITH NANOMATERIALS

11.6.1 Electronic devices

Printed electronics use printing methods to build electrical circuits and devices (Kamyshny and Magdassi, 2019). Producing electrical devices on rigid substrates using planar technologies like photolithography or vacuum deposition is standard practice. These procedures have two flaws; firstly, subtractive manufacturing technologies are primarily planar making enclosed geometries, voids, overhangs, and 3D objects difficult to fabricate (Lin et al., 2019, Lai et al., 2019). Secondly, most traditional technologies can only process one material per stage. This inhibits their potential to build versatile devices. However, using 3D printing of nanomaterials enables us to customize and produce multi-material and multi-scaled devices. Using nanoparticles in electronic printing can improve electrical conductivity and optical characteristics. Nanomaterials' functional qualities have been used to create highly efficient energy storage devices including microsupercapacitors and lithium-ion batteries, as well as flexible electronics and photonic devices (Ferris et al., 2019).

11.6.2 Energy storage devices

Flexible and wearable devices require high-performance micro-scale energy storage devices with high rates of charging/discharging and high energy density (Sun et al., 2013). Batteries store energy by electrochemical reactions between two electrodes: cathode and anode. Many nanomaterials have been employed to fabricate electrolytes and electrodes used in energy storage devices. Nanomaterial-based energy devices that are produced by 3D printing such as pseudocapacitors, microsupercapacitors, and lithium-ion batteries have recently become possible. Assembling energy storage devices could be simplified if additive manufacturing could print different materials simultaneously (Torres-Canas et al., 2019). Unlike traditional subtractive processes, 3D printing technology permits the inclusion of numerous functional nanomaterials. Also, the assembly of other components with nanomaterials in an electrochemical device could be tuned for preserving their unique qualities like electrical conductivity and high surface area. Printing energy storage devices made of polymer nanocomposites is possible due to the low temperature required for 3D printing technologies like stereolithography. Polymer printing in a dielectric layer is crucial for energy storage devices like capacitors because it increases polarization (Dai et al., 2017).

11.6.3 Flexible electronics

Flexible electronics are devices that can withstand mechanical twisting and bending while their functional features like electrical conductivity are preserved (Zou et al., 2018). Graphene and carbon nanotubes have been used to achieve this flexibility. Pressure sensors, energy harvesting devices, opto-electronic devices, and electronic skin have all been researched in recent decades. The creation of electrical circuits on flexible and soft substrates like polyamides and paper is possible through 3D printing (Lu et al., 2018). Paper can be used to make gadgets that are light, cheap, and biodegradable. Making flexible devices from natural ingredients is another benefit of 3D printing using nanomaterials. Three-dimensional printing's capacity to manufacture complicated geometries may enable textile-based electronics. Electronics have been incorporated into textiles for wearable applications (known also as e-textiles). E-textiles are porous, soft, breathable, and lightweight, all desirable qualities for a wearable device (Gonçalves et al., 2018, Liao et al., 2018).

11.6.4 Photonic devices

Photonic devices also benefit from 3D printing nanomaterials. Light-emitting diodes (LEDs) and photovoltaic and solar cells along with displays are examples of photonic devices that produce, detect, or manipulate light. Photonic devices often use semiconducting nanomaterials as the active layer, such as quantum dots (QDs), perovskite crystals, and quantum dot nanocomposites nanocomposites (Howard et al., 2019). Semiconductor QDs, for example, have optoelectronic capabilities governed by size-dependent quantum confinement phenomena, such as photoluminescence emission spectra. Changing their size allows for tuning their optoelectronic properties (Bao et al., 2015, Asadi et al., 2019).

Three-dimensional printing technology may be used to fabricate LEDs for display applications by patterning semiconductor nanomaterials. A QD with controllable emission wavelength, high luminescent efficiency, and narrow emission spectrum is desirable in such applications. There is extreme compatibility between QDs in colloidal solution and extrusion-based 3D printing methods. LEDs are preferred for display applications due to their color purity and high brightness, while with traditional displays such as monitors and smartphones, their maximum value reaches 600 cd m^{-2}. Achieving these features requires carefully placing semiconductor nanoparticles in a 3D-printed construct (Yang et al., 2019).

Three-dimensional printing and nanomaterials may improve solar cell performance. Some solar cells produced by 3D printing have recently outperformed spin-coated equivalents. Three-dimensional printing can also modify the crystallization of perovskite layers, which is important for solar cell lifetime and performance.

11.7 BIOMEDICAL DEVICES AND THEIR CUSTOMIZATION

Biomedical equipment is any device designed for medical use. They help diagnose, monitor, and cure diseases, and also support important functions.

Biomedical devices include surgical sutures, stents, and pacemakers. The biomedical device sector is large, with a 409.5 billion USD market by 2023. With the use of 3D printing, the invention of biomedical devices that can address various unmet clinical requirements has been enabled (Ventola, 2014). To name a few applications, 3D printing has facilitated the development of implants, constructs for tissue engineering, prosthetics, and biosensors. Three-dimensional printing has dramatically increased the potential to construct innovative biomedical equipment. Incorporating nanoparticles into 3D printing is a promising way to modify mechanical, biological, geometric, or response aspects of biomedical devices (Ten Kate et al., 2017, Dekker et al., 2018, Zamani et al., 2021).

11.7.1 Mechanical properties

Tensile strength, elasticity, and stiffness are key mechanical parameters for biomedical device developments. For example, mechanical characteristics similar to biological tissues have long been an aim of tissue engineering (Niinomi and Nakai, 2011). Another field has focused on designing specific implants whose Young's modulus is modified so that they match native tissues' mechanical properties. The stress shielding effect occurs when an implant's mechanical strength exceeds that of bone, causing bone resorption and increased fracture risk (Sezer and Eren, 2019). Implants having mechanical qualities similar to bone are desired to address this issue; therefore, it is required to engineer medical devices' mechanical properties.

The mechanical properties of 3D-printed biomedical devices can be modified by nanomaterials. Nanomaterials can also be employed to reinforce 3D-printed hydrogels. Because of their biocompatibility, hydrogels have long been researched as tissue scaffolds. The addition of nanoparticles to 3D-printed hydrogels increases the use of these scaffolds in biomedical implants. Three-dimensional printing with nanomaterials also improves device flexibility. Some devices require flexibility to be capable of functioning within the dynamic body. Three-dimensional printing of nanomaterials helps mimic native tissue mechanical characteristics. Using magnetic particles in 3D printing allows for more control over mechanical properties, which is important for biomimetic mechanical properties (Xu et al., 2019, Martin et al., 2015).

11.7.2 Geometric properties

Additive manufacturing can create novel geometries whose production are not possible with the use of subtractive or formative manufacturing (Lee et al., 2014). Medical devices are interested in the ability to produce precise geometries because they can determine numerous functional aspects. Personalized anatomic models can also be created using computer imaging like computerized tomography (CT) scans and 3D printing. These models make the design of individualized medical devices possible and nanomaterials add to 3D printing versatility in these areas (Xu et al., 2016). It can create new materials, personalized medicinal devices, as well as high-resolution structures. Incorporating nanoparticles allows for a greater selection of materials and hence more complex design. Moreover, nanomaterial 3D printing can create customized support systems. To ensure optimal patient healthcare, personalized devices could be used in broader populations (Naftulin et al., 2015).

11.7.3 Biological properties

Initially, bioinert materials were used to promote biocompatibility. These devices can induce certain biological reactions. This could be used in regenerative medicine. Implantable scaffolds are synthesized in order to promote cell differentiation and proliferation while restoring tissue function (Ghosh et al., 2018). Three-dimensional printed nanomaterials can trigger biological processes, resemble biological tissues, and boost cell viability. For example, 3D printing with nanomaterials could produce constructs that imitate biological structures. Three-dimensional bioprinting was used to create an artificial meniscus with seeded human adipose-derived stem cells (Narayanan et al., 2016). The scaffold was printed in the shape of a human meniscus from PLA nanofiber-alginate hydrogel. 3D-printed nanofibers can also be used to create tendon and ligament tissue scaffolds. Natural healing of tendon and ligament injuries results in scar-like tissue with poor performance. Current synthetic or allograft implants have limited biocompatibility, longevity, and functioning (Pajala et al., 2009).

Three-dimensional printing using nanoparticles could also be used to fabricate microchannels that mimic vascular structures. In fact, vascularization is still a major obstacle to efficient three-dimensional tissue regeneration. As an example, calvaria bone was made of tricalcium phosphate (TCP) nanoparticle-doped polycaprolactone (PCL). This structure improves vascularization after five months. Similar techniques can be used to increase the biocompatibility of various tissue scaffolds (Kang et al., 2016). Three-dimensional printing of nanomaterials can also limit bacterial growth on the surfaces of printed objects, which is considered a serious risk in implantable medical devices. It was reported that PLA/silver nanowire nanocomposites could render effective protection of implants against *E. coli* and *S. aureus* bacteria. These approaches may reduce the infection risk of medical implants by coating their surfaces with bactericidal nanocomposites (Bayraktar et al., 2019).

11.7.4 Responsive properties

Four-dimensional printing that allows 3D printed materials to change shape after manufacturing, has several potential biomedical uses. Four-dimensional-printed materials and gadgets are responsive to particular stimuli. To give a

3D-printed object a 4D effect, smart materials are used, which can undergo shape change when exposed to external stimuli (Kuang et al., 2019). To provide 4D properties, nanomaterials can be used in 3D printing. A force or external stimuli, such as electrical actuation or optical illumination, can cause a shape or construct change in a nanomaterial. Graphene, silica, and carbon nanotubes can be used to render 4D effects. Polymers are smart materials for 3D-printed biological systems since they are biocompatible and printable. During printing, a permanent structure is formed by polymer crosslinking. After printing, the material can be shaped temporarily. After then, it will revert to its former shape if the temperature exceeds its glass transition temperature. With this capability, structures can move in reaction to stimuli (Narayanan et al., 2016).

Temperature-responsive nanomaterials can give 3D-printed items 4D characteristics. Due to the fact that mechanical characteristics are largely dependent on material temperature, dynamic mechanical properties are possible. For example, at higher temperatures, PNIPAM/silica nanoparticle composites show a solid-to-rubber transition (Guo et al., 2019).

Light-responsive drug delivery systems can be manufactured by 3D printing of nanomaterials. By photothermally heating materials, light-responsive nanoparticles can change shape (Lin et al., 2018). Light is a desirable stimulation source due to its excellent tunability and precision (Jang et al., 2020). In this regard, polycaprolactone nanofibers coated with photothermally active polypyrrole were electrospun and after four exposures to near-infrared light (NIR), up to 60% of the drug was released. Furthermore, printed objects with responsive properties can be produced by water-induced swelling. Hydrogel composites can develop 4D architectures with the capability to twist and fold when exposed to water (Sydney Gladman et al., 2016).

Three-dimensional magnetic nanoparticles can produce 4D characteristics. Magnetic-responsive materials, like light-responsive materials, change temperature when stimulated. The frequency change in an alternating magnetic field around the device causes movement among the magnetic nanoparticles leading to an increase in the temperature (Niiyama et al., 2018).

11.8 BIOELECTRONICS

Electricity in bodily function involves electrical impulses conveyed from the brain through nerves to every organism's component, as well as electrical signals regulating the activity of a cell or group of cells (Ostroverkhova, 2013). Understanding how biological electricity works and how it regulates physiological functions might help us understand how the biological system works. Bioelectronics is the merging of electronics and biology. Luigi Galvani pioneered the field in the 1970s by electrically actuating a frog limb (Zhang and Lieber, 2016). Keeping electronics functioning in a complex and dynamic biological environment is one of the fundamental issues in most bioelectronics (Someya et al., 2016). Traditional electronics are particularly difficult to integrate with biological structures because of mechanical, material, and geometrical barriers. For example, Young's moduli of biological organs are three to six orders of magnitude lower than microfabricated electronics. To interface seamlessly with non-planar complex, biological geometry, technologies must be similar in size to biological cells and tissues (Kong et al., 2016).

Using 3D printing and nanomaterials, devices can be highly tuned in terms of functional properties. For instance, multi-scale additive manufacturing makes it simple to print electronic elements in complex shapes with flexible and biocompatible materials. The approach has a unique capacity to combine biological and electronic systems. Three-dimensional printing bioelectronic scaffolds synthesized based on nanomaterials could help with tissue regeneration and this multi-scale method can produce difficult-to-make nanomaterial-based lab-on-a-chip devices and biosensors (Katz and Willner, 2004).

11.8.1 Microelectrodes

Microelectrodes are able to interact with the neurological pathways in the body to acquire or stimulate electrical impulses. Microelectrodes must be chemically inert and mechanically compliant in order to gain the highest biocompatibility in a complex biological environment (Khodagholy et al., 2011). Nanomaterial-based microelectrode arrays (MEA) are possible to be made faster, more customized, more biocompatible, and cheaper than standard devices. Three-dimensional printing technique can minimize MEA production time and allows for extensive customization. Compared to a single microelectrode, MEAs have better mass transfer, reaction times, sensitivity, and lower limit of detection (LOD). Using silver nanoparticle ink and polymer traces, Yang et al. created MEAs (Yang et al., 2016). Comparing 3D printing to micromachining, they found that 3D printing allows for more precise control of trace spacing at 30 µm, 100 µm, and 180 µm. Kundu et al. established a technique that allowed entire 8 × 8 MEAs to be constructed in under four days (Kundu et al., 2019). The finished object is made from an SLA-printed clear base, ink-cast silver nanoparticle, pulsed electroplating of Pt or Au solution, and laser micromachining. The use of 3D printing processes allows for rapid production and prototyping, allowing for faster deployment into personalized medical devices and structures.

11.8.2 Bioelectronics scaffolds

The combination of bioregenerative scaffolds and electrical components can revolutionize the design of biomedical devices. Electrostimulation, for example, has been offered as a viable technique to enhance tissue regeneration. Electronics embedded in biological scaffolds may augment, modulate, or facilitate regenerative processes. The electrical component integration with biological constructions is attractive but has yet to be achieved because of the mentioned constraints. The integration of electronics into complex regenerative systems may be improved by 3D printing of nanomaterial-based devices (Qazi et al., 2014, Rajabi et al., 2015).

A multiscale 3D printing technique can integrate electrically conductive devices into bioprinted constructions boosting cell adherence along with cell proliferation over passive scaffolds. Chen et al., for example, investigated the impact of aligned nanofibers by seeding mouse skeletal muscle tissue (myoblasts) on electrospun nanofibers of poly(e-caprolactone)(PCL) and PCL/ polyaniline (PANi) (Chen et al., 2013). In a study by Bolin et al., good adhesion and proliferation of neural cells were observed on conductive poly(3,4-ethylenedioxythiophene) (PEDOT)-coated electrospun nanofibers. Electrical stimulation increased intracellular Ca2+ signaling (Bolin et al., 2009).

11.8.3 Biosensors

Biosensors detect the presence of a chemical and respond with an electrical signal. Not all biosensors communicate with electrical impulses of the body; however, biosensors and bioelectronics are treated together because of being closely intertwined. Glucose biosensors, for example, are bioelectronic due to employing electrodes to determine the number of electron transfers during a catalyzed glucose reaction (Malhotra, 2017, Zhang and Liu, 2016).

The addition of nanomaterials into 3D printed structures enables the creation of customizable chemosensors by leveraging their size-dependent functional features. It has been established that 3D-printed nanomaterial-based sensors are capable of detecting compounds such as herbicides, biomarkers, glucose, and bacteria (Su and Chen, 2018).

Three-dimensional printing can also exploit the chemical sensitivity and biocompatibility of metallic nanoparticles to produce flexible electrode arrays that can detect cancer biomarkers and antioxidants. Its accuracy and multi-material flexibility allow complicated electrode arrays to be swiftly built with minimum materials and tuned to the responses of sensors. Three-dimensional printing's quick prototyping capability optimizes biosensor performance in a variety of assay conditions. Computer-aided design can sort bacteria and filter complex combinations to maximize the efficacy of the nanomaterial-based biosensor (Ko et al., 2021, Carvajal et al., 2018).

11.8.4 Lab on a chip

Microfluidic chips or miniaturized arrays and sensors, known also as bio-chips, are examples of lab-on-a-chip research. Labs on a chip are often made using multi-step lithographic methods. Rapid customization and improved cell monitoring could be enabled by 3D printing lab-on-a-chip devices. This enables quick modification besides enhanced biocompatibility. To create a dissolved oxygen sensor for a liver-on-a-chip, Moya et al. used inkjet printing. For sensitive substrates, they used gold and silver nanoparticle inkjet printing at low temperatures. The sensors were flexible, thin, and biocompatible, blending flawlessly with the device's microfluidics and liver cell culture. The microfluidic channels were not 3D printed; however, this could allow for faster production and more flexibility (Moya et al., 2018).

Cell monitoring is improved by 3D-printed nanoparticles. The use of nanoparticles as markers in DNA sensors, oxygen, and pH sensors, as well as immunoassays has been reported. Trampe et al. showed one such nanoparticle sensor by bioprinting living cells together with oxygen-sensing nanoparticles (Trampe et al., 2018). Luminescent nanoparticles of styrene maleic anhydride copolymer, containing O2 luminescent indicator, were 3D printed by the DIW method with mesenchymal stem cells and green microalgae in basic scaffolds, enabling online monitoring of the cells' metabolic activities via a simple single-lens reflex camera. Rapid customization of lab-on-a-chip devices and biosensors, combining diverse cell cultures and indications into a single printed device is feasible by 3D printing (Singh et al., 2015, Zhou et al., 2016, Park et al., 2018).

11.9 POSSIBLE PROBLEMS AND SOLUTION

Three-dimensional bioprinting is a fast-evolving process that uses biomaterial inks and precise microstructures to control cell movement and function. Due to the ability of this technique to precisely and reliably design and generate unique spatial structures, it is possible to fabricate 3D complex bone constructs for tissue regeneration in plastic surgery, dentistry, and orthopedics (Dhawan et al., 2019). Topological microstructures, nanocomponents of biomaterial inks, cell sources, and bioactive factors or drugs are critical to successfully produce bone tissues and consequently affect the shape fidelity, multilevel architecture, mechanical properties, and biological functions of the bone regeneration implants (Derakhshanfar et al., 2018).

Three-dimensional printing and bioprinting of bone structures have been made possible by mixing structural features, cells, and functional biomaterial inks. Inductive and conductive nanoparticles could also be used to recreate differentiation microenvironments and vascular networks. A synergistic combination can control the destiny of stem cells or progenitor cells in created tissues. But several critical concerns remain unresolved. In order to 3D print bone constructs with gradient transition structures from nano- to micro-scales, shape stability, and high mechanical properties, researchers should have a better combination of the advantages of nanotechnology and nanomaterials such as nanofibers, nanoparticles, nanocrystals, and nanosheets. Second, it is required to regulate osteogenesis and angiogenesis of the embedded stem cells or recruited progenitor cells by nanomaterials, bioactive medicines or chemicals, and nanotopography signals provided via the bioprinted constructions. Finally, the gradual alteration in shape and physicochemical properties of nanoscale materials or their in vivo degradation while implanted in the body is a crucial issue to achieve long-term in vivo safety (Masaeli et al., 2019). There is a lack of sufficient studies to assess the biosafety of tailored implants, particularly in terms of nanomaterial toxicity and aggregation. In order to establish an osteogenic microenvironment equivalent to the complicated natural bone-related tissues, it is required to have an in-depth study of nanomaterials, nanotechnology, and 3D printing or bioprinting (Lyons et al., 2020).

11.10 SUMMARY AND OUTLOOK

This chapter discussed the incorporation of nanostructures into 3D printing. Adding nanomaterials can affect dimensional accuracy, lower sintering temperatures increase thermal and electric conductivity, and enhance mechanical properties.

Despite these early results, using nanomaterials in 3D printing poses many obstacles. When using nanoparticles with 3D printing media, each approach has unique constraints (aggregation within printing media, nozzle clogging, the rough surface texture of produced items, etc.). To successfully 3D print with nanomaterials, new instruments may be required to overcome process and material boundaries. There are considerable information gaps regarding the use of nanomaterials in 3D printing. For example, several nanomaterials have yet to be used in 3D printing. Also, little is known about how nanocomposites interact with printing media. There are no standard parameters or synthesis methods for distinct nanomaterials or processes. These issues require substantial attention. The combination of 3D printing and nanotechnology opens up numerous new possibilities. Nanostructures' adjustable qualities allow us to broaden the material properties and thereby applications of printed parts. Additionally, by altering the nanostructure loadings during synthesis, 3D-printed nanocomposites may be able to construct things with graded material properties. Combining different nanomaterials in a single AM part would allow us to print more sophisticated structures like fuel cells, batteries, solar cells, and so on. Nanobiomaterials developments may enable the printing of replacement organs and bone. These future paths may entail new materials for existing 3D printing technologies and entirely new AM processes. To fully utilize the promise of nanomaterials and AM, much more research is required. Researchers have more work to do, but the rewards might be substantial.

REFERENCES

Abuchenari, A. & Moradi, M. 2019. The Effect of Cu-substitution on the micro-structure and magnetic properties of Fe-15% Ni alloy prepared by mechanical alloying. *Journal of Composites and Compounds*, 1, 10–15.

Al-Milaji, K. N., Secondo, R. R., Ng, T. N., Kinsey, N. & Zhao, H. 2018. Interfacial Self-Assembly of Colloidal Nanoparticles in Dual-Droplet Inkjet Printing. *Advanced Materials Interfaces*, 5, 1701561.

Al Abadi, H., Thai, H.-T., Paton-Cole, V. & Patel, V. 2018. Elastic properties of 3D printed fibre-reinforced structures. *Composite Structures*, 193, 8–18.

Araya-Calvo, M., López-Gómez, I., Chamberlain-Simon, N., León-Salazar, J. L., Guillén-Girón, T., Corrales-Cordero, J. S. & Sánchez-Brenes, O. 2018. Evaluation of compressive and flexural properties of continuous fiber fabrication additive manufacturing technology. *Additive Manufacturing*, 22, 157–164.

Asadi, E., Chimeh, A. F., Hosseini, S., Rahimi, S., Sarkhosh, B., Bazli, L., Bashiri, R. & Tahmorsati, A. H. V. 2019. A review of clinical applications of graphene quantum dot-based composites. *Journal of Composites and Compounds*, 1, 31–40.

Azadani, R. N., Sabbagh, M., Salehi, H., Cheshmi, A., Raza, A., Kumari, B. & Erabi, G. 2021. Sol-gel: Uncomplicated, routine and affordable synthesis procedure for utilization of composites in drug delivery. *Journal of Composites and Compounds*, 3, 57–70.

Bao, B., Li, M., Li, Y., Jiang, J., Gu, Z., Zhang, X., Jiang, L. & Song, Y. 2015. Patterning fluorescent quantum dot nanocomposites by reactive inkjet printing. *Small*, 11, 1649–1654.

Bayraktar, I., Doganay, D., Coskun, S., Kaynak, C., Akca, G. & Unalan, H. E. 2019. 3D printed antibacterial silver nanowire/polylactide nanocomposites. *Composites Part B: Engineering*, 172, 671–678.

Boley, J. W., Chaudhary, K., Ober, T. J., Khorasaninejad, M., Chen, W. T., Hanson, E., Kulkarni, A., Oh, J., Kim, J. & Aagesen, L. K. 2017. High-Operating-Temperature Direct Ink Writing of Mesoscale Eutectic Architectures. *Advanced Materials*, 29, 1604778.

Bolin, M. H., Svennersten, K., Wang, X., Chronakis, I. S., Richter-Dahlfors, A., Jager, E. W. & Berggren, M. 2009. Nano-fiber scaffold electrodes based on PEDOT for cell stimulation. *Sensors and Actuators B: Chemical*, 142, 451–456.

Bückmann, T., Stenger, N., Kadic, M., Kaschke, J., Frölich, A., Kennerknecht, T., Eberl, C., Thiel, M. & Wegener, M. 2012. Tailored 3D mechanical meta-materials made by dip-in direct-laser-writing optical lithography. *Advanced Materials*, 24, 2710–2714.

Buffat, P. & Borel, J. P. 1976. Size effect on the melting temperature of gold particles. *Physical Review A*, 13, 2287.

Carvajal, S., Fera, S. N., Jones, A. L., Baldo, T. A., Mosa, I. M., Rusling, J. F. & Krause, C. E. 2018. Disposable inkjet-printed electrochemical platform for detection of clinically relevant HER-2 breast cancer biomarker. *Biosensors and Bioelectronics*, 104, 158–162.

Castro, J. M. & Rajaraman, S. 2022. Experimental and Modeling Based Investigations of Process Parameters on a Novel, 3D Printed and Self-Insulated 24-Well, High-Throughput 3D Microelectrode Array Device for Biological Applications. *Journal of Microelectromechanical Systems*.

Cerkvenik, U., Van De Straat, B., Gussekloo, S. W. & Van Leeuwen, J. L. 2017. Mechanisms of ovipositor insertion and steering of a parasitic wasp. *Proceedings of the National Academy of Sciences*, 114, E7822–E7831.

Chabaud, G., Castro, M., Denoual, C. & L. E. Duigou, A. 2019. Hygromechanical properties of 3D printed continuous carbon and glass fibre reinforced polyamide composite for outdoor structural applications. *Additive Manufacturing*, 26, 94–105.

Chakraborty, S. & Biswas, M. C. 2020. 3D printing technology of polymer-fiber composites in textile and fashion industry: A potential roadmap of concept to consumer. *Composite Structures*, 248, 112562.

Chavez, L. A., Regis, J. E., Delfin, L. C., Garcia Rosales, C. A., Kim, H., Love, N., Liu, Y. & Lin, Y. 2019. Electrical and mechanical tuning of 3D printed photopolymer–MWCNT nanocomposites through in situ dispersion. *Journal of Applied Polymer Science*, 136, 47600.

Chen, M.-C., Sun, Y.-C. & Chen, Y.-H. 2013. Electrically conductive nanofibers with highly oriented structures and their potential application in skeletal muscle tissue engineering. *Acta Biomaterialia*, 9, 5562–5572.

Crane, N. B., Wilkes, J., Sachs, E. & Allen, S. M. 2006. Improving accuracy of powder-based SFF processes by metal deposition from a nanoparticle dispersion. *Rapid Prototyping Journal*.

Dai, X., Deng, Y., Peng, X. & Jin, Y. 2017. Quantum-dot light-emitting diodes for large-area displays: towards the dawn of commercialization. *Advanced Materials*, 29, 1607022.

Dekker, T. J., Steele, J. R., Federer, A. E., Hamid, K. S. & Adams Jr, S. B. 2018. Use of patient-specific 3D-printed titanium implants for complex foot and ankle limb salvage, deformity correction, and arthrodesis procedures. *Foot & Ankle International*, 39, 916–921.

Derakhshanfar, S., Mbeleck, R., Xu, K., Zhang, X., Zhong, W. & Xing, M. 2018. 3D bioprinting for biomedical devices and tissue engineering: A review of recent trends and advances. *Bioactive Materials*, 3, 144–156.

Dhawan, A., Kennedy, P. M., Rizk, E. B. & Ozbolat, I. T. 2019. Three-dimensional bioprinting for bone and cartilage restoration in orthopaedic surgery. *JAAOS-Journal of the American Academy of Orthopaedic Surgeons*, 27, e215–e226.

Duan, B., Wang, M., Zhou, W. Y., Cheung, W. L., Li, Z. Y. & Lu, W. W. 2010. Three-dimensional nanocomposite scaffolds fabricated via selective laser sintering for bone tissue engineering. *Acta Biomaterialia*, 6, 4495–4505.

Eskandarinezhad, S., Khosravi, R., Amarzadeh, M., Mondal, P. & Magalhães Filho, F. J. C. 2021. Application of different Nanocatalysts in industrial effluent treatment: A review. *Journal of Composites and Compounds*, 3, 43–56.

Ferris, A., Bourrier, D., Garbarino, S., Guay, D. & Pech, D. 2019. 3D interdigitated microsupercapacitors with record areal cell capacitance. *Small*, 15, 1901224.

Gao, Q., Niu, X., Shao, L., Zhou, L., Lin, Z., Sun, A., Fu, J., Chen, Z., Hu, J. & Liu, Y. 2019. 3D printing of complex GelMA-based scaffolds with nanoclay. *Biofabrication*, 11, 035006.

Ghosh, U., Ning, S., Wang, Y. & Kong, Y. L. 2018. Addressing unmet clinical needs with 3D printing technologies. *Advanced Healthcare Materials*, 7, 1800417.

Gissibl, T., Thiele, S., Herkommer, A. & Giessen, H. 2016. Two-photon direct laser writing of ultracompact multi-lens objectives. *Nature Photonics*, 10, 554–560.

Goesmann, H. & Feldmann, C. 2010. Nanoparticulate functional materials. *Angewandte Chemie International Edition*, 49, 1362–1395.

Gonçalves, C., Ferreira Da Silva, A., Gomes, J. & Simoes, R. 2018. Wearable e-textile technologies: A review on sensors, actuators and control elements. *Inventions*, 3, 14.

González Flores, G. A., Bertana, V., Chiappone, A., Roppolo, I., Scaltrito, L., Marasso, S. L., Cocuzza, M., Massaglia, G., Quaglio, M., Pirri, C. F. & Ferrero, S. 2022. Single-Step 3D Printing of Silver-Patterned Polymeric Devices for Bacteria Proliferation Control. *Macromolecular Materials and Engineering*, 307, 2100596.

Goodridge, R. D., Shofner, M. L., Hague, R. J., Mcclelland, M., Schlea, M., Johnson, R. & Tuck, C. J. 2011. Processing of a Polyamide-12/carbon nanofibre composite by laser sintering. *Polymer Testing*, 30, 94–100.

Greenhall, J., Guevara Vasquez, F. & Raeymaekers, B. 2016. Ultrasound directed self-assembly of user-specified patterns of nanoparticles dispersed in a fluid medium. *Applied Physics Letters*, 108, 103103.

Guo, Y., Belgodere, J. A., Ma, Y., Jung, J. P. & Bharti, B. 2019. Directed Printing and Reconfiguration of Thermoresponsive Silica-pNIPAM Nanocomposites. *Macromolecular Rapid Communications*, 40, 1900191.

Hales, S., Tokita, E., Neupane, R., Ghosh, U., Elder, B., Wirthlin, D. & Kong, Y. L. 2020. 3D printed nanomaterial-based electronic, biomedical, and bioelectronic devices. *Nanotechnology*, 31, 172001.

Hetrick, D. R., Sanei, S. H. R., Bakis, C. E. & Ashour, O. 2021. Evaluating the effect of variable fiber content on mechanical properties of additively manufactured continuous carbon fiber composites. *Journal of Reinforced Plastics and Composites*, 40, 365–377.

Howard, I. A., Abzieher, T., Hossain, I. M., Eggers, H., Schackmar, F., Ternes, S., Richards, B. S., Lemmer, U. & Paetzold, U. W. 2019. Coated and printed perovskites for photovoltaic applications. *Advanced Materials*, 31, 1806702.

Huang, L., Du, X., Fan, S., Yang, G., Shao, H., Li, D., Cao, C., Zhu, Y., Zhu, M. & Zhang, Y. 2019. Bacterial cellulose nanofibers promote stress and fidelity of 3D-printed silk based hydrogel scaffold with hierarchical pores. *Carbohydrate Polymers*, 221, 146–156.

Huang, Y., Tian, X., Zheng, Z., Li, D., Malakhov, A. V. & Polilov, A. N. 2022. Multiscale concurrent design and 3D printing of continuous fiber reinforced thermoplastic composites with optimized fiber trajectory and topological structure. *Composite Structures*, 285, 115241.

Jang, H. W., Zareidoost, A., Moradi, M., Abuchenari, A., Bakhtiari, A., Pouriamanesh, R., Malekpouri, B., Rad, A. J. & Rahban, D. 2020. Photosensitive nanocomposites: environmental and biological applications. *Journal of Composites and Compounds*, 2, 50–60.

Joo, H. & Cho, S. 2020. Comparative studies on polyurethane composites filled with polyaniline and graphene for DLP-type 3D printing. *Polymers*, 12, 67.

Kabir, S., Mathur, K. & Seyam, A.-F. M. 2020. The road to improved fiber-reinforced 3D printing technology. *Technologies*, 8, 51.

Kabir, S. F., Mathur, K. & Seyam, A.-F. M. 2020. A critical review on 3D printed continuous fiber-reinforced composites: History, mechanism, materials and properties. *Composite Structures*, 232, 111476.

Kamyshny, A. & Magdassi, S. 2019. Conductive nanomaterials for 2D and 3D printed flexible electronics. *Chemical Society Reviews*, 48, 1712–1740.

Kang, H.-W., Lee, S. J., Ko, I. K., Kengla, C., Yoo, J. J. & Atala, A. 2016. A 3D bioprinting system to produce human-scale tissue constructs with structural integrity. *Nature Biotechnology*, 34, 312–319.

Katz, E. & Willner, I. 2004. Biomolecule-functionalized carbon nanotubes: applications in nanobioelectronics. *ChemPhysChem*, 5, 1084–1104.

Kazemzadeh, A., Meshkat, M. A., Kazemzadeh, H., Moradi, M., Bahrami, R. & Pouriamanesh, R. 2019. Preparation of graphene nanolayers through surfactant-assisted pure shear milling method. *Journal of Composites and Compounds*, 1, 22–26.

Khodagholy, D., Doublet, T., Gurfinkel, M., Quilichini, P., Ismailova, E., Leleux, P., Herve, T., Sanaur, S., Bernard, C. & Malliaras, G. G. 2011. Highly conformable conducting polymer electrodes for in vivo recordings. *Advanced Materials*, 23, H268–H272.

Ko, W.-Y., Huang, L.-T. & Lin, K.-J. 2021. Green technique solvent-free fabrication of silver nanoparticle–carbon nanotube flexible films for wearable sensors. *Sensors and Actuators A: Physical*, 317, 112437.

Kong, Y. L., Gupta, M. K., Johnson, B. N. & Mcalpine, M. C. 2016. 3D printed bionic nanodevices. *Nano Today*, 11, 330–350.

Kuang, X., Roach, D. J., Wu, J., Hamel, C. M., Ding, Z., Wang, T., Dunn, M. L. & Qi, H. J. 2019. Advances in 4D printing: materials and applications. *Advanced Functional Materials*, 29, 1805290.

Kundu, A., Nattoo, C., Fremgen, S., Springer, S., Ausaf, T. & Rajaraman, S. 2019. Optimization of makerspace microfabrication techniques and materials for the realization of planar, 3D printed microelectrode arrays in under four days. *RSC Advances*, 9, 8949–8963.

Lai, X., Guo, R., Lan, J., Geng, L., Lin, S., Jiang, S., Zhang, Y., Xiao, H. & Xiang, C. 2019. Flexible reduced graphene oxide/electroless copper plated poly (benzo)-benzimidazole fibers with electrical conductivity and corrosion resistance. *Journal of Materials Science: Materials in Electronics*, 30, 1984–1992.

Lee, J.-S., Hong, J. M., Jung, J. W., Shim, J.-H., Oh, J.-H. & Cho, D.-W. 2014. 3D printing of composite tissue with complex shape applied to ear regeneration. *Biofabrication*, 6, 024103.

Li, J., Hill, E. H., Lin, L. & Zheng, Y. 2019. Optical nanoprinting of colloidal particles and functional structures. *ACS nano*, 13, 3783–3795.

Liao, X., Song, W., Zhang, X., Huang, H., Wang, Y. & Zheng, Y. 2018. Directly printed wearable electronic sensing textiles towards human–machine interfaces. *Journal of Materials Chemistry C*, 6, 12841–12848.

Lin, D., Jin, S., Zhang, F., Wang, C., Wang, Y., Zhou, C. & Cheng, G. J. 2015. 3D stereolithography printing of graphene oxide reinforced complex architectures. *Nanotechnology*, 26, 434003.

Lin, Q., Li, L., Tang, M., Hou, X. & Ke, C. 2018. Rapid macroscale shape morphing of 3D-printed polyrotaxane monoliths amplified from pH-controlled nanoscale ring motions. *Journal of Materials Chemistry C*, 6, 11956–11960.

Lin, X., Wu, M., Zhang, L. & Wang, D. 2019. Superior stretchable conductors by electroless plating of copper on knitted fabrics. *ACS Applied Electronic Materials*, 1, 397–406.

Lu, B., Lan, H. & Liu, H. 2018. Additive manufacturing frontier: 3D printing electronics. *Opto-Electronic Advances*, 1, 170004.

Lyons, J. G., Plantz, M. A., Hsu, W. K., Hsu, E. L. & Minardi, S. 2020. Nanostructured biomaterials for bone regeneration. *Frontiers in Bioengineering and Biotechnology*, 922.

Madou, M. J. 2018. *Fundamentals of Microfabrication and Nanotechnology, Three-volume Set*, CRC Press.

Malhotra, B. D. 2017. *Biosensors: Fundamentals and Applications*, Smithers rapra.

Martin, J. J., Fiore, B. E. & Erb, R. M. 2015. Designing bioinspired composite reinforcement architectures via 3D magnetic printing. *Nature Communications*, 6, 1–7.

Masaeli, R., Zandsalimi, K., Rasoulianboroujeni, M. & Tayebi, L. 2019. Challenges in three-dimensional printing of bone substitutes. *Tissue Engineering Part B: Reviews*, 25, 387–397.

Moya, A., Ortega-Ribera, M., Guimerà, X., Sowade, E., Zea, M., Illa, X., Ramon, E., Villa, R., Gracia-Sancho, J. & Gabriel, G. 2018. Online oxygen monitoring using integrated inkjet-printed sensors in a liver-on-a-chip system. *Lab on a Chip*, 18, 2023–2035.

Naftulin, J. S., Kimchi, E. Y. & Cash, S. S. 2015. Streamlined, inexpensive 3D printing of the brain and skull. *PloS one*, 10, e0136198.

Narayanan, L. K., Huebner, P., Fisher, M. B., Spang, J. T., Starly, B. & Shirwaiker, R. A. 2016. 3D-bioprinting of polylactic acid (PLA) nanofiber–alginate hydrogel bioink containing human adipose-derived stem cells. *ACS biomaterials Science & Engineering*, 2, 1732–1742.

Naseer, N., Bashir, S., Latief, N., Latif, F., Khan, S. N. & Riazuddin, S. J. R. M. 2018. Human amniotic membrane as differentiating matrix for in vitro chondrogenesis. 13, 821–832.

Niazvand, F., Cheshmi, A., Zand, M., Nasrazadani, R., Kumari, B., Raza, A. & Nasibi, S. 2020. An overview of the development of composites containing Mg and Zn for drug delivery. *Journal of Composites and Compounds*, 2, 193–204.

Niinomi, M. & Nakai, M. 2011. Titanium-based biomaterials for preventing stress shielding between implant devices and bone. *International Journal of Biomaterials*, 2011.

Niiyama, E., Uto, K., Lee, C. M., Sakura, K. & Ebara, M. 2018. Alternating magnetic field-triggered switchable nanofiber mesh for cancer thermo-chemotherapy. *Polymers*, 10, 1018.

Ostroverkhova, O. 2013. *Handbook of Organic Materials for Optical and (Opto) Electronic Devices: Properties and Applications*, Elsevier.

Pajala, A., Kangas, J., Siira, P., Ohtonen, P. & Leppilahti, J. 2009. Augmented compared with nonaugmented surgical repair of a fresh total Achilles tendon rupture: a prospective randomized study. *JBJS*, 91, 1092–1100.

Park, C. S., Ha, T. H., Kim, M., Raja, N., Yun, H.-S., Sung, M. J., Kwon, O. S., Yoon, H. & Lee, C.-S. 2018. Fast and sensitive near-infrared fluorescent probes for ALP detection and 3d printed calcium phosphate scaffold imaging in vivo. *Biosensors and Bioelectronics*, 105, 151–158.

Qazi, T. H., Rai, R. & Boccaccini, A. R. 2014. Tissue engineering of electrically responsive tissues using polyaniline based polymers: A review. *Biomaterials*, 35, 9068–9086.

Rajabi, A. H., Jaffe, M. & Arinzeh, T. L. 2015. Piezoelectric materials for tissue regeneration: A review. *Acta Biomaterialia*, 24, 12–23.

Reiser, A., Lindén, M., Rohner, P., Marchand, A., Galinski, H., Sologubenko, A. S., Wheeler, J. M., Zenobi, R., Poulikakos, D. & Spolenak, R. 2019. Multi-metal electrohydrodynamic redox 3D printing at the submicron scale. *Nature Communications*, 10, 1–8.

Reyes, C., Fu, L., Suthanthiraraj, P. P., Owens, C. E., Shields IV, C. W., López, G. P., Charbonneau, P. & Wiley, B. J. 2018. The Limits of Primary Radiation Forces in Bulk Acoustic Standing Waves for Concentrating Nanoparticles. *Particle & Particle Systems Characterization*, 35, 1700470.

Rong, Y., Hu, Y., Mei, A., Tan, H., Saidaminov, M. I., Seok, S. I., Mcgehee, M. D., Sargent, E. H. & Han, H. 2018. Challenges for commercializing perovskite solar cells. *Science*, 361, eaat8235.

Saleh, A. H., Kumar, D., Sirakov, I., Shafiee, P. & Arefian, M. 2021. Application of nano compounds for the prevention, diagnosis, and treatment of SARS-coronavirus: A review. *Journal of Composites and Compounds*, 3, 230–246.

Sezer, H. K. & Eren, O. 2019. FDM 3D printing of MWCNT re-inforced ABS nanocomposite parts with enhanced mechanical and electrical properties. *Journal of Manufacturing Processes*, 37, 339–347.

Shin, D., Kim, J., Choi, S., Lee, Y.-B. & Chang, J. 2019. Droplet-jet mode near-field electrospinning for controlled helix patterns with sub-10 μm coiling diameter. *Journal of Micromechanics and Microengineering*, 29, 045004.

Singh, H., Shimojima, M., Shiratori, T., Van An, L., Sugamata, M. & Yang, M. 2015. Application of 3D printing technology in increasing the diagnostic performance of enzyme-linked immunosorbent assay (ELISA) for infectious diseases. *Sensors*, 15, 16503–16515.

Someya, T., Bao, Z. & Malliaras, G. G. 2016. The rise of plastic bioelectronics. *Nature*, 540, 379–385.

Su, C.-K. & Chen, J.-C. 2018. One-step three-dimensional printing of enzyme/substrate–incorporated devices for glucose testing. *Analytica Chimica Acta*, 1036, 133–140.

Sun, K., Wei, T. S., Ahn, B. Y., Seo, J. Y., Dillon, S. J. & Lewis, J. A. 2013. 3D printing of interdigitated Li-Ion microbattery architectures. *Advanced Materials*, 25, 4539–4543.

Suvanasuthi, R., Chimnaronk, S. & Promptmas, C. 2022. 3D printed hydrophobic barriers in a paper-based biosensor for point-of-care detection of dengue virus serotypes. *Talanta*, 237, 122962.

Sydney Gladman, A., Matsumoto, E. A., Nuzzo, R. G., Mahadevan, L. & Lewis, J. A. 2016. Biomimetic 4D printing. *Nature Materials*, 15, 413–418.

Tan, A. T., Beroz, J., Kolle, M. & Hart, A. J. 2018. Direct-write freeform colloidal assembly. *Advanced Materials*, 30, 1803620.

Ten Kate, J., Smit, G. & Breedveld, P. 2017. 3D-printed upper limb prostheses: a review. *Disability and Rehabilitation: Assistive Technology*, 12, 300–314.

Torres-Canas, F., Yuan, J., Ly, I., Neri, W., Colin, A. & Poulin, P. 2019. Inkjet printing of latex-based high-energy microcapacitors. *Advanced Functional Materials*, 29, 1901884.

Trampe, E., Koren, K., Akkineni, A. R., Senwitz, C., Krujatz, F., Lode, A., Gelinsky, M. & Kühl, M. 2018. Functionalized Bioink with Optical Sensor Nanoparticles for O2 Imaging in 3D-Bioprinted Constructs. *Advanced Functional Materials*, 28, 1804411.

Ventola, C. L. 2014. Medical applications for 3D printing: current and projected uses. *Pharmacy and Therapeutics*, 39, 704.

Wang, Q., Phung, N., Di Girolamo, D., Vivo, P. & Abate, A. 2019. Enhancement in lifespan of halide perovskite solar cells. *Energy & Environmental Science*, 12, 865–886.

Xu, N., Wei, F., Liu, X., Jiang, L., Cai, H., Li, Z., Yu, M., Wu, F. & Liu, Z. 2016. Reconstruction of the upper cervical spine using a personalized 3D-printed vertebral body in an adolescent with Ewing sarcoma. *Spine*, 41, E50–E54.

Xu, W., Molino, B. Z., Cheng, F., Molino, P. J., Yue, Z., Su, D., Wang, X., Willför, S., Xu, C. & Wallace, G. G. 2019. On low-concentration inks formulated by nanocellulose assisted with gelatin methacrylate (GelMA) for 3D printing toward wound healing application. *ACS Applied Materials & Interfaces*, 11, 8838–8848.

Yang, H., Rahman, M. T., Du, D., Panat, R. & Lin, Y. 2016. 3-D printed adjustable microelectrode arrays for electrochemical sensing and biosensing. *Sensors and Actuators B: Chemical*, 230, 600–606.

Yang, Z., Gao, M., Wu, W., Yang, X., Sun, X. W., Zhang, J., Wang, H.-C., Liu, R.-S., Han, C.-Y. & Yang, H. 2019. Recent advances in quantum dot-based light-emitting devices: Challenges and possible solutions. *Materials Today*, 24, 69–93.

Yang, Z., Yang, Z., Chen, H. & Yan, W. 2022. 3D printing of short fiber reinforced composites via material extrusion: Fiber breakage. *Additive Manufacturing*, 58, 103067.

Yim, S., Goyal, K. & Sitti, M. 2013. Magnetically actuated soft capsule with the multimodal drug release function. *IEEE/ASME Transactions on Mechatronics*, 18, 1413–1418.

Yim, S. & Sitti, M. 2012. Shape-programmable soft capsule robots for semi-implantable drug delivery. *IEEE Transactions on Robotics*, 28, 1198–1202.

Yugang, D., Yuan, Z., Yiping, T. & Dichen, L. 2011. Nano-TiO2-modified photosensitive resin for RP. *Rapid Prototyping Journal*.

Yunus, D. E., Shi, W., Sohrabi, S. & Liu, Y. 2016. Shear induced alignment of short nanofibers in 3D printed polymer composites. *Nanotechnology*, 27, 495302.

Zamani, Y., Ghazanfari, H., Erabi, G., Moghanian, A., Fakić, B., Hosseini, S. M. & Mahammod, B. P. 2021. A review of additive manufacturing of Mg-based alloys and composite implants. *Journal of Composites and Compounds*, 3, 71–83.

Zamani, Y., Zareein, A., Bazli, L., Nasrazadani, R., Mahammod, B. P., Nasibi, S. & Chahardehi, A. M. 2020. Nanodiamond-containing composites for tissue scaffolds and surgical implants: A review. *Journal of Composites and Compounds*, 2, 215–227.

Zhai, X., Ma, Y., Hou, C., Gao, F., Zhang, Y., Ruan, C., Pan, H., Lu, W. & Liu, W. 2017. 3D-printed high strength bioactive supramolecular polymer/clay nanocomposite hydrogel scaffold for bone regeneration. *ACS Biomaterials Science & Engineering*, 3(6), 1109–1118.

Zhang, A. & Lieber, C. M. 2016. Nano-bioelectronics. *Chemical Reviews*, 116, 215–257.

Zhang, D. & Liu, Q. 2016. Biosensors and bioelectronics on smartphone for portable biochemical detection. *Biosensors and Bioelectronics*, 75, 273–284.

Zhang, Q., Zhang, F., Xu, X., Zhou, C. & Lin, D. 2018. Three-dimensional printing hollow polymer template-mediated graphene lattices with tailorable architectures and multifunctional properties. *ACS Nano*, 12, 1096–1106.

Zhang, Y., Zhang, F., Yan, Z., Ma, Q., Li, X., Huang, Y. & Rogers, J. A. 2017. Printing, folding and assembly methods for forming 3D mesostructures in advanced materials. *Nature Reviews Materials*, 2, 1–17.

Zhou, X., Zhu, W., Nowicki, M., Miao, S., Cui, H., Holmes, B., Glazer, R. I. & Zhang, L. G. 2016. 3D bioprinting a cell-laden bone matrix for breast cancer metastasis study. *ACS Applied Materials & Interfaces*, 8, 30017–30026.

Zou, M., Ma, Y., Yuan, X., Hu, Y., Liu, J. & Jin, Z. 2018. Flexible devices: from materials, architectures to applications. *Journal of Semiconductors*, 39, 011010.

Chapter 12

4D Printing of Smart Magnetic-Based Robotic Materials

Ali Zolfagharian, Mir Irfan Ul Haq, Marwan Nafea, and Mahdi Bodaghi

CONTENTS

12.1 INTRODUCTION

Robots may be seen in a new context as something that shares a collaborative environment with humans for medical and assistance means. This cannot be realized by conventional rigid robots and manufacturing (Whitesides, 2018; Zolfagharian et al., 2020a, 2016a). This is one of the future themes, concentrating on factors such as robot flexibility and intelligence embodiment for adaptive interaction, compliance, and safety recently achieved via soft robotics. This sort of robotics relies heavily on the use of smart materials. But to deliver proper smart functionalities of the future robots, one could optimize the controlled movement and force using smart robotic materials and four-dimensional (4D) printing.

Advancements in smart material functionality and additive manufacturing enable programmable modulations in size, form, and stiffness that allow sophisticated soft robots to be actuated by external stimuli of electric, magnetic, chemical, thermal, or photonic sources, expanding the possibilities of 4D printing of soft robotics.

DOI: 10.1201/9781003306238-12

Responsive materials that can change their size and/or stiffness in response to electrical and magnetic fields are the essence of electromagnetically powered soft robots. These materials must be able to endure mechanical stress while maintaining a low elastic modulus.

Magnetism may be used to direct the soft actuator's translational and rotational motion in a variety of ways. Elastomers and hydrogels are often mixed with micro- or nano-particles (Yarali et al., 2022). In elastomeric composites, the dispersion of local electromagnetic properties may enhance motion control precision. Continuously distributed magnetization has been established to form sections with arbitrary magnetic sections that may be used to regulate motion (Zhang and Diller, 2018). These techniques can be used to generate novel robotic materials for soft actuators using 3D printing, origami, and kirigami (Xu et al., 2017; Kim et al., 2018).

The electromagnetic smart robotic materials suitable for 4D printing are categorized in this chapter, from the material point of view, into magnetorheological elastomers (MREs), magnetic-based shape memory polymers (MSMPs), and magnetic-based hydrogels.

12.2 MAGNETORHEOLOGICAL ELASTOMERS IN ROBOTIC 4D PRINTING

A magnetic particle suspension in a viscous liquid is represented by an MRE, which is a physical representation of an MR fluid. Seismic vibration dampers and artificial joints are all examples of robotic applications where MREs are often employed (Chung et al., 2021). The ordering force of magnetostatic particle-particle interaction, in MREs, outweighs the randomizing force of particle thermal motion, resulting in chain-like particle aggregates oriented along the direction of the magnetic field. An increase in the MR fluid's modulus is caused by the creation of chain-like aggregates, which stiffens the fluid.

12.2.1 Magnetic particle dispersion strategies in MREs

Optimal characteristics of magnetic elastomers and hydrogels are achieved when the magnetic particles are evenly dispersed without aggregation. Blending, in situ precipitation and grafting are the most popular procedures for dispersing magnetic particles (Li et al., 2013). The blending, which is called physical doping, involves combining magnetic particles, uncrosslinked polymer chains, and crosslinking agents in a single container that is usually filled with a solvent or nonpolymerizing silicone oil (Chung et al., 2021). During the crosslinking process, the magnetic particles, which do not form chemical connections with polymeric components, are held in place. In spite of its simplicity, the blending process still has limitations with dispersion and particle agglomeration control. Weak chemical bonds between the particles and the polymeric network make bleaching magnetic particles in swelling gels more difficult.

Hydrogels, but not elastomers, may be suitable for using in situ precipitation. A precursor solution containing magnetic metals is infused into a polymeric gel network, and then a chemical reaction is triggered to precipitate magnetic particles in the gel. The in-situ precipitation approach produces more uniformly dispersed magnetic particles than the blending method, making it easier to include a large number of particles. This approach, on the other hand, is only acceptable for hydrogels that are able to withstand alkalines (Zhao et al., 2015).

Magnetic elastomers are often made via the grafting-onto approach, in which the magnetic particle functions as a crosslinker by possessing chemically functional groups on its surface. The magnetic composite material made via this approach is more stable because of the covalent connection between the particles and the polymer chains (Evans et al., 2012).

Magnetic particles may have an isotropic or anisotropic distribution. While it is simpler to make isotropic materials, a magnetic field may be used to linearly organize the particles while the elastomer matrix is curing. The actuation behavior may be given directionality by such an arrangement of magnetic particles (Chung et al., 2021). Magnetic actuation may be directed and efficient if the magnetic particles are trapped in microsized domains (Majidi and Wood, 2010). Only particular areas of the actuator are affected by an applied field, due to the distribution of magnetic particles inside predefined regions. By 3D printing patterns encased in nonmagnetic silicone, the effects of design and orientation on the efficacy of vibration damping in relation to applied magnetic field magnitude and direction were studied (Bastola et al., 2017).

To 4D print MREs soft robots, the ferromagnetic particles are dispersed inside silicon- or rubber-based polymers (such as PDMS) that serve as excellent electrical insulators, such as iron, cobalt, or neodymium–iron–boron (NdFeB) (Li et al., 2020). The external magnetic field also aligns the randomly oriented particles, resulting in changes in stiffness and damping (Bastola et al., 2020) and actuations. Magnetic fields may improve the stiffness and shear modulus of MREs based on the particle size, concentration, and magnetic properties as well as the elastomer matrix (Banerjee and Ren, 2018; Xu et al., 2019). In the presence of a magnetic field, experiments have shown that randomly distributed magnetic particles exhibit less stiffness variation than anisotropic MREs (Ruddy, et al., 2012). Also, it was found that the volume fraction of the magnetic particles could be varied to find the optimum value for the maximum stiffness increase in soft manipulators with the introduction of the magnetic field (Zhang et al., 2008).

Particle concentration and MREs soft robots compliance were shown to be linearly inversely related (Fuhrer et al., 2013). Therefore, the design of soft robotic MREs requires a trade-off between stiffness amplification and desired conformance. Magnetic particle alignment and the degree of crosslinking affect the deflection and mechanical properties of MRE soft actuators (Lee et al., 2020; Tian and Nakano, 2018). To see how well a magnetic soft actuator held up under different magnetic fields and filler concentrations, a gripper was developed and tested (Tian and Nakano, 2018). The results of the experiments revealed the direct relationship between the soft gripper bending and the magnetic filler concentration up to a certain level of the magnetic field, 80 mT. After this level, the bending of the gripper remained quite stable. The gripping force of the magnetic soft gripper, however, was reported to be quite low, around 0.1 N, in response to the magnetic field of 80 mT.

To manage stress concentration and adhesion in deformed robotic bodies, MREs have been used as sensing components in soft robotics. Highly stretchable MREs made from a mixture of Ecoflex 00–10 (Smooth-On, PA, USA) and iron powder were integrated into a soft pneumatic actuator as an inductive coil-based wireless sensing mechanism to measure external loads and internal pressure in the precision of micrometres' deformation of the soft robot body (Wang et al., 2019). The developed MREs sensor showed the capability of measurement of actuator's deformation with no hysteresis caused by either internal pressure or external contacts. Liquid metal-filled MRE was also developed as a sensing module exhibiting maximum resistivity in the rest state while it is dropping drastically under the mechanical deformations (Yun et al., 2019).

12.2.2 Electro-magnetorheological elastomers in robotic 4D printing

In comparison to other soft robotics research, the electro-magnetorheological elastomers (EMREs) have received significantly less attention. Composites of metal and/or magnetic particles and elastomers having electromagnetic properties are widely used to build soft actuators based on EMREs principles, which respond to electrical stimulation in the presence of a magnetic field.

In research of soft robotics, a Ni nanowire/silicone nanocomposite actuator with electro-magnetic properties was tested in response to the magnetic force direction (Park et al., 2008). Increased concentrations of metal particles, according to the findings, led to greater actuation strain due to improved electrical responsiveness (Park et al., 2008). In addition, the strain response of EMREs soft actuators dropped with reasonable phase lag when the input voltage frequency was increased (Park et al., 2008). The EMRE soft actuators have demonstrated manageable hysteresis (Park et al., 2008). However, they do not respond effectively to low voltage in high frequencies (Zhu et al., 2019). Active vibration control of soft robots might potentially benefit from the EMREs soft actuators. It is, however, the inherent time delays in these systems must be addressed.

Soft robots that use electrorheological fluids with dispersed magnetic particles and electric and magnetic fields for actuation modulation are also developed (Bastola et al., 2018; Sadeghi et al., 2012). MR fluid was injected into the 3D-printed hollow struts in a field-responsive 3D-printed metamaterial architecture (Figure 12.1(a)) (Jackson et al., 2018). The product displayed dynamically regulated movement under an applied magnetic field. Slightly quicker soft robots, however, still face hurdles when using low-viscosity electrorheological fluids. With EMREs in soft robotics, sedimentation and homogeneous redistribution of magnetic particles are also significant practical difficulties.

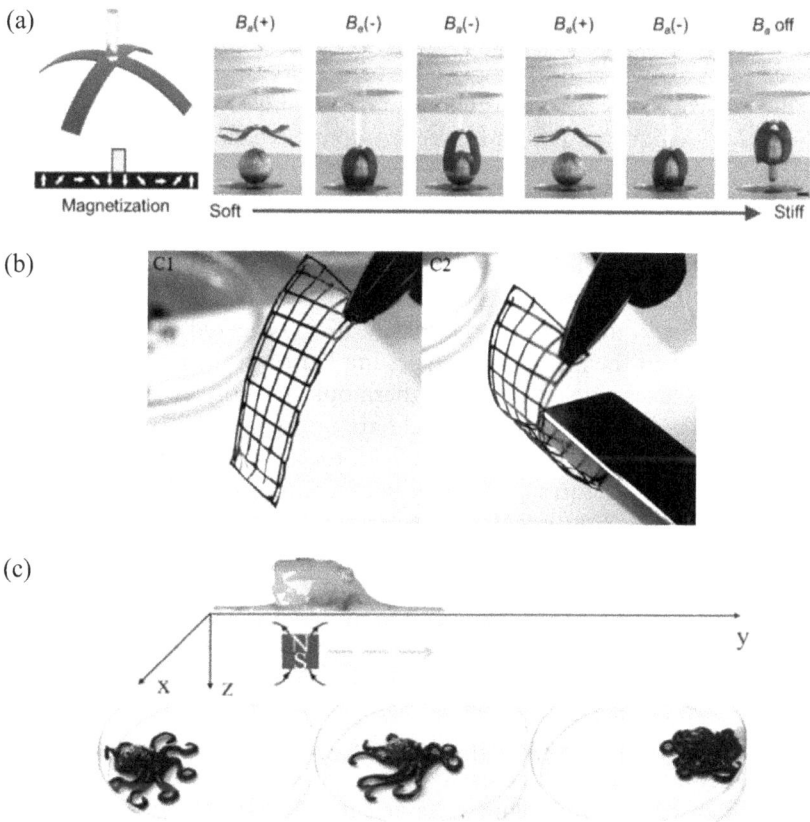

Figure 12.1 (a) A magnetic shape-locking SMP 4D-printed with and without locking mechanisms (left and right). (b) The soft neodymium magnet hydrogel responsivity to the magnetic field that can be folded and twisted. (c) A 4D-printed magnetic-responsive octopus hydrogel soft robot.

12.3 MAGNETIC-BASED SHAPE MEMORY POLYMERS IN ROBOTIC 4D PRINTING

There has been a recent surge in the use of SMPs in the manufacturing of soft robotics. Heat-responsive SMPs (Bodaghi et al., 2016; Zolfagharian et al., 2018a) have been the focus of the bulk of studies, although this is not always ideal since it is difficult to induce locally controlled deformation via the thermal environment. As a result, it is critical to have SMPs that are acceptably sensitive to stimulus and have more local controllability. There has been more recent research on the creation of magnetic-responsive SMPs in 4D printing of soft robots (Kim et al., 2018; Shao et al., 2020).

With its capacity to regulate signal direction and strength, as well as its simplicity in penetration into polymers, magnetic stimulation was a good choice for polymer and hydrogel-based soft robotics (Lum et al., 2016). Medicinal robots, such as crawling and swimming ones, have profited from such remote and high-frequency control capabilities (Liu et al., 2019).

In biomedical applications, such as catheters, an SMP-based magnetically actuated variable stiffness soft robot (Liu et al., 2019) with simultaneous magnetic actuation and photothermal heating is practical (Chautems et al., 2017). In order to lower the modulus of SMP and enable actuation with applied magnetic fields, photothermal heating triggered by an LED was employed to heat the magnetic particles implanted in the material. In order to make the magnetic SMP sheet, thermoplastic polyurethane polymers were used. The LED and magnetic field are utilised to regulate the stiffness of SMP. In the absence of a magnetic field, the actuator is reshaped into its final state by simply turning the light back on.

To generate motion in polymer-based soft robots, magnetic fillers are often patterned onto an elastomeric layer. Applying a magnetic field aligns the fillers resulting in a variety of actuation modes, including bending and twisting. Despite the current application of magnetic soft robots, there are certain drawbacks to their use since the coils they use to generate external magnetic fields are heavy and demand a lot of power, making them burdensome in a small area.

Since magnetic micro- and nano-particles may be used to create new types of reconfigurable soft robots using a variety of manufacturing techniques, including casting, electrospinning, laser cutting, and 3D printing, this has opened the door to a new class of 4D printing magnetic-based soft robots (Liu et al., 2019; Zhang et al., 2015; Mohr et al., 2006; Wei et al., 2017). Polylactic acid (PLA) was 3D-printed as a magnetic responsive soft actuator with the addition of Fe_3O_4 nanoparticles (Wei et al., 2017). The prepared ink was UV cured after extrusion from the printer nozzle. The developed soft robots showed a reversible remote actuation to the alternating magnetic field. Magnetic SMP soft robots also have a locking mechanism that allows for rapid and easy system reconfiguration (Figure 12.1(a)). To do so, two different types of filler particles were integrated to induce actuation and local heating independently (Ze et al., 2020).

12.4 MAGNETIC-BASED HYDROGELS IN ROBOTIC 4D PRINTING

Another class of soft and elastic materials is hydrogels, which are hydrophilic polymers capable of absorbing large amounts of water. New materials for flexible polymer printing include hydrogels. Hydrogels may be used to make soft robotics devices provided their mechanical properties allow for actuation without rupturing the structures (Zolfagharian et al., 2016b).

One of the hydrogel's most notable features is that it may be used in a variety of water-based applications. Because of their high viscosity and the range of printing methods available, hydrogels are well-suited for application in the 4D printing of soft robotics (Shiblee et al., 2019; Zolfagharian et al., 2018b). Hydrogels that respond to stimuli like electrical and magnetic stimuli are increasingly used in biomedical soft robotics because their reversible shape-shifting may be swiftly activated. As a result of their high water content, hydrogels are useful for a wide range of biomedical applications, including drug delivery (Dong et al., 2020; Shi et al., 2019) and tissue engineering (Champeau et al., 2020) soft robots.

Hydrogel-based soft robots use volumetric changes in swelling and shrinkage as their primary actuation mechanism. Multiple methods exist, including bi-material (Zolfagharian et al., 2019) and multi-material (Zhai et al., 2020; Zolfagharian et al., 2020b) composite, materials patterning (Erb et al., 2013), and functionally graded crosslinks (Lin et al., 2019) to achieve desired movements such as bending and twists. The polarity and amplitude of the external magnetic field control the actuation direction and bending deflection of the hydrogel in which magnetic nanoparticles (MNPs), such as Fe_2O_3, Fe_3O_4, and $CoFe_2O_4$, are disseminated into the precursor hydrogel (Zolfagharian et al., 2016b; Gao et al., 2016). Yet, in order to avoid particle agglomeration, the dispersion medium for nanoparticles must be sufficiently viscous (Podstawczyk et al., 2020). Soft robots of this sort have found use in the pharmaceutical and medication delivery industry, where the controlled and non-invasive release of medicinal dosages is possible (Figure 12.1(b)).

Microparticles of NdFeB alloy and fumed silica nanoparticles were mixed with a silicone rubber matrix for the preparation of auxetic-type soft robotic hydrogel. To increase the printability of the ink via shear thinning as well as shape fidelity after extrusion the fumed silica was introduced as a rheological modifier. The prepared elastomer matrix was magnetized under the impulse magnetic field in the curation stage (Kim et al., 2018). This study demonstrated the application of magnetic responsive hydrogel soft robots with fast-moving metamaterial characteristics in pharmaceutical applications.

Direct three-dimensional (3D) printing has lately been used to manufacture complex form magnetic-based robots with unique properties such as functionally graded and variable stiffness, which can be controlled remotely (Kim et al., 2018; Podstawczyk et al., 2020; Kim et al., 2020, 2019).

An octopus soft robot was developed by dispersing ferromagnetic Fe_3O_4 nanoparticles in polyacrylamide (PAAm) (Chen et al., 2019) demonstrating a quick response to the magnetic stimulus (Figure 12.1(c)). Encapsulating Fe_3O_4 nanoparticles inside an alginate ionic hydrogel created a hydrogel soft actuator that could be manufactured in 4D (McCracken et al., 2019). The new actuator has been tested in aqueous soft robotics applications. Yet in practical applications, hydrogels' mechanical strength is the most important consideration. With nanocomposite inclusion, the mechanical strength of pure composites with electrical and magnetic fields was improved (Lee et al., 2019).

A 3D printing ink was created based on a combination of alginate, methylcellulose, polyacrylic acid (PAA), and Fe_3O_4 MNPs with the capability of developing convolutional geometry while being responsive to the magnetic field (Podstawczyk et al., 2020). The study has improved the rheological properties of the ink for direct 3D printing. Also, it was demonstrated that the infill pattern made via 3D printing led to a controlled bending of magnetic responsive soft hydrogel actuators.

12.5 DISCUSSIONS OF SMART MAGNETIC-BASED ROBOTIC MATERIALS IN 4D PRINTING

MR elastomers differ from MR liquids in that they can transform morphology considerably in addition to their capacity to adjust viscosity (Sun et al., 2017). A variety of novel properties, including 3D printability (Qi et al., 2019) and lower range viscosity, anisotropy for directional actuation (Zhang et al., 2020), surface-modified magnetic particles for improved dispersion (Zhou et al., 2020), novel polymer matrix for reduced cytotoxicity (Cvek et al., 2020), and fiber reinforcement of MR elastomers for matrix enhancement and magnetic particle alignment (Shu et al., 2020) are investigated in MREs. These materials will enable previously unimaginable material properties tailored with additive manufacturing. Peristaltic pumps (Wu et al., 2019) and soft grippers (Shu et al., 2020) are only some of the recent examples of MREs (Alapan et al., 2020). All these applications will be further investigated using 4D printing of magnetic soft actuators used in soft robotics.

EMREs have been determined to be the least studied of all the electric and magnetic-responsive polymer soft robots evaluated in this research. As a result, further research into the uses of electrorheological fluids in soft robotics should focus on the size, dispersion, and kind of magnetic particles, as well as the optimal viscosity.

Using MREs, small-scale soft robots made of hard magnetic particles encapsulated in silicone were constructed to function under very low

Reynolds circumstances (Zhang and Diller, 2018; Wu et al., 2020; Hu et al., 2018). Untethered soft robotics have considerably profited from these features. Various additive manufacturing, including extrusion (Kim et al., 2018; Qi et al., 2020) and voxel-level drop-on-demand 3D printing (Sundaram et al., 2019), were used to construct MREs soft robots.

Four-dimensional-printed magnetic-based soft robots have issues due to their large magnetic field generator and significant energy consumption (Table 12.1). To lower the requisite field amplitude, small magnetic soft robots are preferable (Goudu et al., 2020). It is also critical that greater attention be paid to developing thin membranes with a low mass density that can deform rapidly at high speeds in these kinds of robots (Zhu et al., 2018). A combination of different stimuli, including electrical, pneumatic, and magnetic (Cao et al., 2019; Liu et al., 2017) response principles have been used to produce modern soft robots. As a consequence of the repulsive magnetic field, the soft robots have shown a significantly enhanced output stroke when activated at the resonance frequency.

When compared to MREs, magnetic SMP soft robots provide greater design freedom since they can be tuned for different glass transition temperature (T_g) values and conductive particle concentrations. Despite this, the locking mechanism of magnetic SMP soft robots is still sluggish. Developing autonomously navigating soft robots constructed of magnetic SMPs might be a promising new research area in the future (Ji et al., 2019; El-Atab et al., 2020).

Further research on hydrogel soft actuators that are magnetic-based might be the future direction in the 4D printing of soft robots. Pioneering research on a freestanding liquid crystalline actuator manufactured from poly(acrylic acid) blocks to stabilize the newly created Fe_3O_4 magnetic microparticles was undertaken (Zhou et al., 2012). Hydro-swelling led to the formation of free-standing liquid crystalline ferrogels because the PAA absorbed a substantial quantity of water while the liquid crystalline domains remained dehydrated and undamaged. In the presence of a magnetic field, the produced ferrogel sample underwent rapid and reversible bending activation.

Hydrogel soft robots, especially in medical applications, have undergone a huge amount of study and accomplishment owing to their bio-compatibility, biodegradability, transparency, and compliance. When it comes to fast and high force throughput in soft robotics, hydrogels have yet to demonstrate their full potential. Synthesis hydrogels that can be produced via 4D printing and generate more force are projected to increase their functioning in the present research areas.

The extrusion-based additive manufacturing of soft materials is a suitable way to print custom hydrogels and commercial silicones. Extrusion-based procedures enable a larger range of materials to be printed with high elongation percentages and heat tolerance, unlike photopolymerization processes and resin-based 3D printing approaches. It has been possible to produce heterogeneous composites with property variation by extruding polymers containing magnetically oriented additive particles (Qi et al., 2020).

Table 12.1 Magnetic-based 4D-printed soft robots

	MREs	EMREs	Magnetic SMPs	Magnetic Hydrogels
Response time	Low	Medium-High	Medium-High	High
Efficiency	Medium-High	Medium	Medium	Low-Medium
Strain	Low	Medium	Medium	High
Stress	Low-Medium	Medium	Medium	Low
3D/4D printing	(Xu et al., 2019; Qi et al., 2020; Sundaram et al., 2019; Zhu et al., 2018; Alapan et al., 2020)	NA	(Wei et al., 2017)	(Kim et al., 2018; Shao et al., 2020; Lin et al., 2019; Podstawczyk et al., 2020; Kim et al., 2020; Chen et al., 2019)
Non-3D printing	(Kim et al., 2018; Li et al., 2020; Lee et al., 2020)	(Park et al., 2008; Bastola et al., 2018; Sadeghi et al., 2012)	(Liu et al., 2019; Ze et al., 2020)	(Goudu et al., 2020; Zhou et al., 2012)
Advantages	Remotely controlled	Little hysteresis	Easy process Locking	Low temperature Low operating voltage
Challenges	Low force External field interference	Sedimentation Redispersion	High voltage	Low strength

REFERENCES

Alapan, Y., et al., *Reprogrammable shape morphing of magnetic soft machines.* Science Advances, 2020. 6(38): p. eabc6414.

Banerjee, H., and H. Ren, *Electromagnetically responsive soft-flexible robots and sensors for biomedical applications and impending challenges,* in Electromagnetic Actuation and Sensing in Medical Robotics. 2018, Springer. pp. 43–72.

Bastola, A., V. Hoang, and L. Li, *A novel hybrid magnetorheological elastomer developed by 3D printing.* Materials & Design, 2017. 114: pp. 391–397.

Bastola, A.K., L. Li, and M. Paudel, *A hybrid magnetorheological elastomer developed by encapsulation of magnetorheological fluid.* Journal of Materials Science, 2018. 53(9): pp. 7004–7016.

Bastola, A.K., et al., *Recent progress of magnetorheological elastomers: a review.* Smart Materials and Structures, 2020.

Bodaghi, M., A. Damanpack, and W. Liao, *Self-expanding/shrinking structures by 4D printing.* Smart Materials and Structures, 2016. 25(10): p. 105034.

Cao, C., X. Gao, and A. Conn, *A compliantly coupled dielectric elastomer actuator using magnetic repulsion.* Applied Physics Letters, 2019. 114(1): p. 011904.

Champeau, M., et al., *4D Printing of Hydrogels: A Review.* Advanced Functional Materials, 2020: p. 1910606.

Chautems, C., et al. *A variable stiffness catheter controlled with an external magnetic field.* in *2017 IEEE/RSJ International Conference on Intelligent Robots and Systems (IROS).* 2017. IEEE.

Chen, Z., et al., *3D Printing of Multifunctional Hydrogels.* Advanced Functional Materials, 2019. 29(20): p. 1900971.

Chung, H.J., A.M. Parsons, and L. Zheng, *Magnetically controlled soft robotics utilizing elastomers and gels in actuation: A review.* Advanced Intelligent Systems, 2021. 3(3): p. 2000186.

Cvek, M., et al., *Poly (2-oxazoline)-based magnetic hydrogels: Synthesis, performance and cytotoxicity.* Colloids and Surfaces B: Biointerfaces, 2020. 190: p. 110912.

Dong, Y., et al., *4D Printed Hydrogels: Fabrication, Materials, and Applications.* Advanced Materials Technologies, 2020: p. 2000034.

El-Atab, N., et al., *Soft Actuators for Soft Robotic Applications: A Review.* Advanced Intelligent Systems, 2020. 2(10): p. 2000128.

Erb, R.M., et al., *Self-shaping composites with programmable bioinspired microstructures.* Nature Communications, 2013. 4(1): pp. 1–8.

Evans, B.A., et al., *A highly tunable silicone-based magnetic elastomer with nanoscale homogeneity.* Journal of Magnetism and Magnetic Materials, 2012. 324(4): pp. 501–507.

Fuhrer, R., et al., *Soft Iron/Silicon Composite Tubes for Magnetic Peristaltic Pumping: Frequency-Dependent Pressure and Volume Flow.* Advanced Functional Materials, 2013. 23(31): pp. 3845–3849.

Gao, W., et al., *Magnetic driving flowerlike soft platform: Biomimetic fabrication and external regulation.* ACS Applied Materials & Interfaces, 2016. 8(22): pp. 14182–14189.

Goudu, S.R., et al., *Biodegradable Untethered Magnetic Hydrogel Milli-Grippers.* Advanced Functional Materials, 2020: p. 2004975.

Hu, W., et al., *Small-scale soft-bodied robot with multimodal locomotion.* Nature, 2018. **554**(7690): pp. 81–85.

Jackson, J.A., et al., *Field responsive mechanical metamaterials.* Science Advances, 2018. **4**(12): p. eaau6419.

Ji, X., et al., *An autonomous untethered fast soft robotic insect driven by low-voltage dielectric elastomer actuators.* Science Robotics, 2019. **4**(37): p.eaaz6451.

Kim, D., et al., *Untethered gripper-type hydrogel microrobot actuated by electric field and magnetic field.* Smart Materials and Structures, 2020.

Kim, Y., et al., *Printing ferromagnetic domains for untethered fast-transforming soft materials.* Nature, 2018. **558**(7709): pp. 274–279.

Kim, Y., et al., *Ferromagnetic soft continuum robots.* Science Robotics, 2019. **4**(33): p. eaax7329.

Lee, J.H., et al., *Highly Tough, Biocompatible, and Magneto-Responsive Fe 3 O 4/ Laponite/PDMAAm Nanocomposite Hydrogels.* Scientific Reports, 2019. **9**(1): pp. 1–13.

Lee, M., et al., *Characterization of a magneto-active membrane actuator comprising hard magnetic particles with varying crosslinking degrees.* Materials & Design, 2020. **195**: p. 108921.

Li, X., et al., *A magneto-active soft gripper with adaptive and controllable motion.* Smart Materials and Structures, 2020.

Li, Y., et al., *Magnetic hydrogels and their potential biomedical applications.* Advanced Functional Materials, 2013. **23**(6): pp. 660–672.

Lin, C., et al., *4D-Printed Biodegradable and Remotely Controllable Shape Memory Occlusion Devices.* Advanced Functional Materials, 2019. **29**(51): p. 1906569.

Liu, J.A.-C., et al., *Photothermally and magnetically controlled reconfiguration of polymer composites for soft robotics.* Science Advances, 2019. **5**(8): p. eaaw2897.

Liu, L., et al., *A biologically inspired artificial muscle based on fiber-reinforced and electropneumatic dielectric elastomers.* Smart Materials and Structures, 2017. **26**(8): p. 085018.

Lum, G.Z., et al., *Shape-programmable magnetic soft matter.* Proceedings of the National Academy of Sciences, 2016. **113**(41): p. E6007–E6015.

Majidi, C., and R.J. Wood, *Tunable elastic stiffness with microconfined magnetorheological domains at low magnetic field.* Applied Physics Letters, 2010. **97**(16): p. 164104.

McCracken, J.M., et al., *Ionic Hydrogels with Biomimetic 4D-Printed Mechanical Gradients: Models for Soft-Bodied Aquatic Organisms.* Advanced Functional Materials, 2019. **29**(28): p. 1806723.

Mohr, R., et al., *Initiation of shape-memory effect by inductive heating of magnetic nanoparticles in thermoplastic polymers.* Proceedings of the National Academy of Sciences, 2006. **103**(10): pp. 3540–3545.

Park, J.-M., et al., *Actuation of electrochemical, electro-magnetic, and electro-active actuators for carbon nanofiber and Ni nanowire reinforced polymer composites.* Composites Part B: Engineering, 2008. **39**(7–8): pp. 1161–1169.

Podstawczyk, D., et al., *3D printed stimuli-responsive magnetic nanoparticle embedded alginate-methylcellulose hydrogel actuators.* Additive Manufacturing, 2020: p. 101275.

Qi, S., et al., *Versatile magnetorheological plastomer with 3D printability, switchable mechanics, shape memory, and self-healing capacity.* Composites Science and Technology, 2019. **183**: p. 107817.

Qi, S., et al., *3D printed shape-programmable magneto-active soft matter for biomimetic applications*. Composites Science and Technology, 2020. **188**: p. 107973.

Ruddy, C., E. Ahearne, and G. Byrne, *A review of magnetorheological elastomers: properties and applications*. Advanced Manufacturing Science (AMS) Research. http://www.ucd.ie/mecheng/ams/news_items/Cillian% 20Ruddy.pdf Accessed, 2012. **20**.

Sadeghi, A., L. Beccai, and B. Mazzolai. *Innovative soft robots based on electro-rheological fluids*. in *2012 IEEE/RSJ International Conference on Intelligent Robots and Systems*. 2012. IEEE.

Shao, L.-H., et al., *4D printing composite with electrically controlled local deformation*. Extreme Mechanics Letters, 2020: p. 100793.

Shi, Q., et al., *Bioactuators based on stimulus-responsive hydrogels and their emerging biomedical applications*. NPG Asia Materials, 2019. **11**(1): pp. 1–21.

Shiblee, M.N.I., et al., *4D printing of shape-memory hydrogels for soft-robotic functions*. Advanced Materials Technologies, 2019. **4**(8): p. 1900071.

Shu, Q., et al., *High performance magnetorheological elastomers strengthened by perpendicularly interacted flax fiber and carbonyl iron chains*. Smart Materials and Structures, 2020. **29**(2): p. 025010.

Sun, S., et al., *Development of an isolator working with magnetorheological elastomers and fluids*. Mechanical Systems and Signal Processing, 2017. **83**: pp. 371–384.

Sundaram, S., et al., *Topology optimization and 3D printing of multimaterial magnetic actuators and displays*. Science Advances, 2019. **5**(7): p. eaaw1160.

Tian, T. and M. Nakano, *Fabrication and characterisation of anisotropic magnetorheological elastomer with 45 iron particle alignment at various silicone oil concentrations*. Journal of Intelligent Material Systems and Structures, 2018. **29**(2): pp. 151–159.

Wang, H., et al. *A wireless inductive sensing technology for soft pneumatic actuators using magnetorheological elastomers*. in *2019 2nd IEEE International Conference on Soft Robotics (RoboSoft)*. 2019. IEEE.

Wei, H., et al., *Direct-write fabrication of 4D active shape-changing structures based on a shape memory polymer and its nanocomposite*. ACS Applied Materials & Interfaces, 2017. **9**(1): pp. 876–883.

Whitesides, G.M., *Soft robotics*. Angewandte Chemie International Edition, 2018. **57**(16): pp. 4258–4273.

Wu, C., et al., *Smart magnetorheological elastomer peristaltic pump*. Journal of Intelligent Material Systems and Structures, 2019. **30**(7): pp. 1084–1093.

Wu, C., et al., *Numerical study of millimeter-scale magnetorheological elastomer robot for undulatory swimming*. Journal of Physics D: Applied Physics, 2020. **53**(23): p. 235402.

Xu, L., T.C. Shyu, and N.A. Kotov, *Origami and kirigami nanocomposites*. Acs Nano, 2017. **11**(8): pp. 7587–7599.

Xu, T., et al., *Millimeter-scale flexible robots with programmable three-dimensional magnetization and motions*. Science Robotics, 2019. **4**(29): p. eaav4494.

Yarali, E., et al., *Magneto-/electro-responsive polymers toward manufacturing, characterization, and biomedical/soft robotic applications*. Applied Materials Today, 2022. **26**: p. 101306.

Yun, G., et al., *Liquid metal-filled magnetorheological elastomer with positive piezoconductivity*. Nature Communications, 2019. **10**(1): pp. 1–9.

Ze, Q., et al., *Magnetic shape memory polymers with integrated multifunctional shape manipulation.* Advanced Materials, 2020. **32**(4): p. 1906657.

Zhai, Y., et al., *Printing Multi-Material Organic Haptic Actuators.* Advanced Materials, 2020: p. 2002541.

Zhang, F., et al., *Remote, fast actuation of programmable multiple shape memory composites by magnetic fields.* Journal of Materials Chemistry C, 2015. **3**(43): p. 11290–11293.

Zhang, J., and E. Diller, *Untethered miniature soft robots: Modeling and design of a millimeter-scale swimming magnetic sheet.* Soft Robotics, 2018. **5**(6): pp. 761–776.

Zhang, J., et al., *The magneto-mechanical properties of off-axis anisotropic magnetorheological elastomers.* Composites Science and Technology, 2020. **191**: p. 108079.

Zhang, X., et al., *Analysis and fabrication of patterned magnetorheological elastomers.* Smart Materials and Structures, 2008. **17**(4): p. 045001.

Zhao, W., et al., *In situ synthesis of magnetic field-responsive hemicellulose hydrogels for drug delivery.* Biomacromolecules, 2015. **16**(8): pp. 2522–2528.

Zhou, Y., et al., *Hierarchically structured free-standing hydrogels with liquid crystalline domains and magnetic nanoparticles as dual physical cross-linkers.* Journal of the American Chemical Society, 2012. **134**(3): pp. 1630–1641.

Zhou, Y., et al., *The fabrication and properties of magnetorheological elastomers employing bio-inspired dopamine modified carbonyl iron particles.* Smart Materials and Structures, 2020. **29**(5): p. 055005.

Zhu, M., et al., *Transient responses of magnetorheological elastomer and isolator under shear mode.* Smart Materials and Structures, 2019. **28**(4): p. 044002.

Zhu, P., et al., *4D printing of complex structures with a fast response time to magnetic stimulus.* ACS Applied Materials & Interfaces, 2018. **10**(42): pp. 36435–36442.

Zolfagharian, A., et al., *Evolution of 3D printed soft actuators.* Sensors and Actuators A: Physical, 2016a. **250**: pp. 258–272.

Zolfagharian, A., et al. *3D printed hydrogel soft actuators.* in *2016 IEEE Region 10 Conference (TENCON).* 2016b. IEEE.

Zolfagharian, A., et al., *Pattern-driven 4D printing.* Sensors and Actuators A: Physical, 2018a. **274**: pp. 231–243.

Zolfagharian, A., et al., *Polyelectrolyte soft actuators: 3D printed chitosan and cast gelatin.* 3D Printing and Additive Manufacturing, 2018b. **5**(2): p. 138–150.

Zolfagharian, A., et al., *Topology-optimized 4D printing of a soft actuator.* Acta Mechanica Solida Sinica, 2019: p. 1–13.

Zolfagharian, A., et al., *Control-based 4D printing: Adaptive 4D-printed systems.* Applied Sciences, 2020a. **10**(9): p. 3020.

Zolfagharian, A., et al., *Effects of Topology Optimization in Multimaterial 3D Bioprinting of Soft Actuators.* International Journal of Bioprinting, 2020b. **6**(2).

Index

For Product Safety Concerns and Information please contact our EU
representative GPSR@taylorandfrancis.com
Taylor & Francis Verlag GmbH, Kaufingerstraße 24, 80331 München, Germany

www.ingramcontent.com/pod-product-compliance
Lightning Source LLC
Chambersburg PA
CBHW060404220326
41598CB00023B/3018

*9 7 8 1 0 3 2 3 0 6 8 2 7 *